# THE
# WEATHER
# MACHINE

ALSO BY ANDREW BLUM
*Tubes*

# THE WEATHER MACHINE

## HOW WE SEE INTO THE FUTURE

# ANDREW BLUM

THE BODLEY HEAD
LONDON

1 3 5 7 9 10 8 6 4 2

The Bodley Head, an imprint of Vintage,
20 Vauxhall Bridge Road,
London SW1V 2SA

The Bodley Head is part of the Penguin Random House group
of companies whose addresses can be found
at global.penguinrandomhouse.com.

 Penguin
Random House
UK

First published in the USA by Ecco in 2019
First published in the UK by The Bodley Head in 2019

www.vintage-books.co.uk

A CIP catalogue record for this book is available from the British Library

Hardback ISBN 9781847923400
Trade paperback ISBN 9781847923417

Printed and bound in Great Britain by Clays Ltd, Elcograf S.p.A.

Penguin Random House is committed to a sustainable future for
our business, our readers and our planet. This book is made
from Forest Stewardship Council® certified paper.

MIX
Paper from
responsible sources
FSC® C018179

TO MICAH & PHOEBE

Perhaps some day in the dim future it will be possible to advance the computations faster than the weather advances and at a cost less than the saving to mankind due to the information gained. But that is a dream.

—Lewis Fry Richardson, 1922

# CONTENTS

# CONTENTS

# THE
# WEATHER
# MACHINE

# PROLOGUE

In October 2012, my son was an infant. I had been counting time carefully, weighing the weeks and days. I was also spending a lot of time on Twitter. I would sit holding him in a rocking chair, the world scrolling by beneath my thumb. It was in that pose, on a Sunday afternoon, that I watched as meteorologists went into a tizzy. The latest run of something called "the European model" had just arrived, and it was sending them into paroxysms. "An organized low-pressure system hasn't even formed yet in the Caribbean, so a LOT can happen," wrote Bryan Norcross, one of the world's most renowned hurricane forecasters, that afternoon. "But because the scenario is so dramatic, it's going to require our attention." The sky outside was clear, and it would be for a week. But the sky on the screen was filled with a storm that didn't yet exist.

Over the next eight days, Superstorm Sandy dumped flooding rains in the Caribbean, headed north across the warm ocean, soaking up energy, then took an extraordinary left turn toward the East Coast, toward New York City, toward us. We pulled down the shades as far as they would go and filled the bathtub with water. The storm came with fury, making the walls restless and twisting the windows in their frames. The lights flickered, and my screen flashed with strange images: the glass carousel on the Brooklyn waterfront floating in the river like a magical barge, downtown streets turned to canals, lampposts sparkling into fireballs. Not far away, the ocean rose up against the land, rushing through living rooms, flooding power stations and corroding the subways' delicate machinery. Neighborhoods along the shore were devastated and Lower Manhattan went dark, a disaster film come to life. At the hospital in which my son had been born, nurses and doctors carried twenty-one infants down unlighted stairways, tangled in battery-powered monitors. Across the region, 147 people died during Sandy, 650,000 homes were damaged or destroyed, and total losses exceeded $50 billion. The city felt fragile. I had the sense that our luck had run out.

New York wasn't the first city to get a storm like this, and it wouldn't be the last. In 2005, Hurricane Katrina startled me not merely with its physical destruction but by the way in which its damage exacerbated inequalities, creating ripple effects throughout society. In 2011, Hurricane Irene gave the Northeast its first experience with what felt like a new kind of storm, whose impacts came less from the wind and more from the rain, which fell for longer periods and in greater quantities than before, raising waters to heights no one could remember. Locally and globally, these storms piled up in a way that made them impossible to ignore. There was scientific debate about the link between these storms and broader climate change, but there was also the bald reality of experience. I had the growing recognition that this was real life now, a new era for earth: record-breaking heat and cold, seasons that stretched in strange directions, weather that in all ways was bigger than before. All as foreseen.

And all as predicted. These storms were different, but so was their anticipation. Something had changed in the weather forecast. The hype was louder and longer, leaving time to pick the bread aisles clean and close schools before the skies had even clouded. Cable television and then social media each created a new and relentless rhythm. But even accounting for that increase in volume, the storms really were bigger, and we certainly knew about them sooner.

The extent of this startled me with Sandy. Norcross's first warning was different not only in degree but in kind. "It's going to require our attention," he wrote a full eight days before, like a forecast for his forecast. His broader concern was, as usual, with the track of the storm and its potential impacts. But his more immediate concern was with the outputs of the computer models. "The most accurate computer forecast models are in amazing agreement today," he pointed out on Sunday. "It's not often that credible forecast models consistently forecast a historic event," he wrote on Tuesday. By Thursday, Norcross was at Defcon 1 and far from alone there. "The strong evidence we have that a significant, maybe historic, storm is going to hit the East Coast is that EVERY reliable computer forecast model now says it's going to happen." Norcross and his colleagues had a view into the evolution of the atmosphere at the spatial scale of hemispheres and the temporal scale of days. This was a long way past merely watching Sandy's development through the space-based camera of a satellite, extrapolating its next move. It was a simulation of the global atmosphere, capable of running ahead of time. Amid a lifetime of weather, it all added up to an improbable, nearly inconceivable, prognostication. I understood that we use computer simulations for weather forecasting. But when had they gotten so good?

In the weeks after Sandy the weather models had a moment of celebrity. They were not new but they were newly powerful. Meteorologists use the word "skill" to judge the accuracy of their predictions, and it has a specific definition: the measure of their ability to forecast the weather better than climatology, meaning the historical average for the place and date. If the average high temperature in New York on March 1 is 45 degrees, any forecast has to be right more often than those climatological averages to count as "skillful." Generally speaking, with each passing decade meteorologists have been able to make that claim one day farther into the future. That means a six-day forecast today is as good as a five-day forecast was a decade ago; a five-day forecast today is as good as a three-day forecast two decades ago; and, most dramatically, today's six-day forecast is as good as a two-day forecast in the 1970s. All of that improvement is thanks to the weather models. It is often credited to "faster supercomputers" or "better satellites." But I suspected it wasn't as simple as that (as if supercomputers and satellites were ever simple). The models were a black box. How did they work? Why were some reliable (and some not)? Who ran them, and who built them? I wanted a look inside.

In my previous book about the physical infrastructure of the Internet—all the data centers and undersea cables and tubes filled with light—I discovered that even the most complex systems are still built by people; they exist in real places and evolve according to some human intention. I learned the most while moving slowly, planting my feet and examining the object in front of me, and talking with the people who built it. I could tell that the source of today's weather forecast was a similar kind of story: complex, ubiquitous and urgent. I knew that if I examined the systems that forecasted the weather with patience and rigor—if I stopped looking up at the sky and instead looked down at the machines that watched it—I might understand this new way of seeing into the future. I wanted to know how the exceptional forecast for Sandy came to be, and what it might tell me about the exceptional forecasts still to come. But I was also curious about the banal, quotidian weather forecasts I looked at every day—like the ones that said it would rain at four o'clock three days from now and often shocked me by being right.

Sandy revealed a paradigm shift in weather forecasting, which now depends less on the day-to-day insights of any human and more on the year-by-year advancements in computer simulations. These prescient weather forecasts were possible not because we had developed a remarkable new skill, but because we had a remarkable new tool. Knowing the weather is one of our oldest desires. After millennia of wishing, we had wired up the earth: with satellites and instrumented balloons; with thermometers, barometers and anemometers; with supercomputers and a purpose-built telecommunications system to tie it all together, in order to see ahead of time.

This global infrastructure of observation and prediction, this weather machine, has many parts and pieces. It has been conceived and constantly improved upon by a group of people few know exist—not the "weathermen" on television but their less visible counterparts: atmospheric scientists, data theorists, satellite makers and diplomats. Notably, it has not been the achievement of a single government agency or corporation but an international construction, a carefully conceived and continuously running system of systems, tuned to an endless loop of observing the weather, predicting the weather, and observing it all over again. The weather machine relies on nearly every major invention of the last three centuries, foremost among them Newtonian physics, telecommunications, spaceflight and computing. It depends on the omnipresent communications system we essentially live inside. It sparkles with computation, of computers' power to consider more variables than any human ever. We touch its technological components every day; it is the umbrella emoji and the forecast high. And we feel its physical analogue in the fresh breeze and the soaking rain.

The weather machine is a wonder we treat as a banality. We look to it every day, turn its outputs into small talk and make judgments about its performance. It marks a high point of science and technology's aspirations for society, but like a lot of things these days, its complex inner workings are not only mysterious but hidden beneath a veneer of simplicity. The forecast is more accurate and more necessary than ever before, while its provenance is harder to discern. We have constructed a tool that we haven't yet learned to trust.

This book is the story of where the weather machine comes from and how it got this way. It is about the protagonists of this superpower: the people who created this window to the future, the people who keep extending its view farther forward in time and the people who might help us better understand the complexity of the world today, in which machines are constantly examining the world, talking to each other and telling us what to do. The ability to forecast the weather is among humanity's greatest adaptations to life on earth. And there was so much to learn about how it all works.

PART I

# CALCULATION

# 1

# Calculating the Weather

On a June afternoon in 2015, I climbed into the old Saab of Anton Eliassen, the director of the Norwegian weather service, known as the Meteorological Institute, and drove up a mountain above Oslo to a restaurant in a century-old wooden lodge. It was springtime in Norway, and the sky was a deep blue stage for passing clouds, each one an aerial mountain range, each one daring a storm. This was a problem. Eliassen had chosen a table for us outside on a broad patio, overlooking the fjord and the harbor. At the brink of seventy, he had a ruddy face and an open manner. He wore a crisp striped dress shirt and had the sharp, attentive ease of a man in charge of his country's weather forecast. A dark cloud was skittering up the hill toward us. Eliassen wrinkled his nose. "It will pass in a minute," he said. When it disappeared over the ridge, leaving us warm and dry and in the sun, Eliassen turned back to his smoked salmon on brown bread. "See? We have excellent short-term forecasting here in Norway."

It was an easy joke for the head of any weather service, but it was particularly pointed in Norway. For a small country, Norway punches above its weight in weather. There were obvious reasons for this: namely, that it is both wealthy and lives more than most at the mercy of the cold and wind. But while most national weather services were founded as part of a naval war department, Norway's was focused on new scientific methods from the start. Professionally and personally, Eliassen is an heir to this tradition. His father, Arnt, had made key contributions to the basic understanding of the mechanics of the atmosphere and worked on the first computer weather models. Anton's boyhood home in Oslo was often filled with famous scientists, arguing at the dinner table or out on the family's sailboat. But they were not drawn by the awesome skies of this "rugged and weathered country above the water," as even Norway's national anthem puts it. They were pencil pushers, not cloud watchers, focused on learning how to predict the weather using math and physics. I pointed this out to Eliassen, and he nodded. "They loved the equations more than the weather," he said.

I did too, I realized. I loved the idea that something as uncontrollable and expansive as the atmosphere could be systematically understood, and that understanding could be so absurdly useful. It was a remarkable leap from mystery to mastery. "The problem is just the application of classical physics to the atmosphere, on a rotating globe, with gravitation," Eliassen said. "That is a beautiful problem, and that is what they were in love with. But it is a rather difficult problem."

How do you calculate the weather? How—before computers, before weather balloons, before satellites—did Norwegian scientists send us down the road toward the computational models we have today?

**When in 1844 Samuel Morse opened the first telegraph line, from Washington to** Baltimore, he famously inquired, "What hath God wrought?" It wasn't meant to be an inquiry about the weather, but from the beginning the telegraph operators seemed to treat it that way. By 1848, there were 2,100 miles of telegraph lines in the United States, but they worked poorly in the rain. When a telegraph operator arrived each morning at the office, he would check in with his counterparts in other cities, to see what the weather was there and prepare for any outages. "If I learned from Cincinnati that the wires to St. Louis were interrupted by rain, I was tolerably sure a 'northeast' storm was approaching," an operator named David Brooks recalled. In Michigan, an operator named Jeptha Homer Wade was known for pinning weather predictions to his bulletin board of "such accuracy as to create considerable comment and wonder." Once the news could travel faster than the winds, then the winds need no longer come as a surprise.

We tend to think of the telegraph as having shrunk the world, "annihilating time and space," in Karl Marx's famous formulation. "Distance and time have been so changed in our imaginations that the globe has been practically reduced in magnitude, and there can be no doubt that our conception of its dimensions is entirely different to that held by our forefathers," mused Josiah Latimer Clark, president of the Society of Telegraph Engineers, presenting an idea that has defined modern life. But when it came to the weather, the telegraph had the opposite effect: It *created* time and space. As soon as information could be exchanged across distances, disparate pieces of sky could be fit together like a puzzle. "The weather" no longer merely described the conditions at a specific place on earth but weather *patterns* that stretched thousands of miles. The weather became bigger than individual experience; it became "a widespread and inter-connected affair, rather than an assortment of local surprises," as James Gleick observed. The weather would no longer merely be the sun or the rain but also a rationally and imaginatively con-structed vision stretching broadly across the land. The weather became a map as much as a breeze.

The art critic and essayist John Ruskin was among the first to recognize what could happen if that weather map could be extended to cover the whole world. Writing in 1839, he imag-ined building "perfect systems of methodical and simultane-ous observations," which he grandly called "a vast machine." Whereas before "the solitary dweller in the American prairie observe the passages of the storms," in the future he "will find himself a part of one mighty Mind—a ray of light entering into one vast Eye." A working telegraph hardly existed at the time of his writing. But Ruskin, while only a twenty-year-old student at Oxford, saw how communication technology would change the way humans imagined not only the weather but the world. "The meteorologist is impotent if alone," he wrote. "His observations are useless; for they are made upon a point, while the specu-lations to be derived from them must be on space." Ruskin's vast machine is equal parts human and technological, powered by cooperation and dependent on communication. He saw that with the telegraph we would no longer be craning our necks to see beyond the horizon; instead, we'd be sailing—at least in the mind's eye—through space, looking down on the winds and the clouds.

The ability to know the weather in many places at one time was the first step toward knowing the weather in one place at many times, most usefully times in the future. Once the telegraph caught on, meteorologists found their work newly practical, and the field was transformed "from weather science to weather service," as the historian James Rodger Fleming has put it. In 1848, the Smithsonian launched a meteorological observation program that aimed to take advantage of the new telegraph networks to provide advance notice of bad weather. When its new headquarters building opened on the Mall in Washington in 1855, the lobby featured a giant map of the United States. Volunteer and paid "Smithsonian Observers," as they were known, would send weather reports in from all over the country, and a paper disk the size of a poker chip was pinned to the location of each report on the lobby map, with conditions indicated by different colors: white for fair weather, black for rain, brown for clouds and blue for snow. "This map is not only of interest to visitors in exhibiting the kind of weather which their friends at a distance are experiencing but is also of importance in determining at a glance the probable changes which may soon be expected," the Smithsonian's directors reported in 1858.

What was being presented was only a general sense of how the weather moved across the country; it came with minimal understanding of how storms formed and evolved. But it was a thrilling attempt at a holistic view, a kind of proto-weather machine. The Smithsonian map could be thought of as an analog predecessor to today's system, like the clerks at early airports who would write arrival and departure times of each flight in chalk. The map pulled in dozens of observations, compared to today's hundreds of millions. But as a leap in possibility, it didn't go unnoticed. The Smithsonian map became a monument, "its daily display of information from all over the country [a] symbol of the way America was growing from a scattering of isolated communities into a single interconnected nation," as the historian Lee Sandlin described it. (That unity cut both ways. When the Civil War severed the telegraph network between North and South, the flow of weather observations stopped and the map was left half blank.)

But that system of observation was not yet organized into a system of prediction. The first routinely distributed forecasts originated soon after in England, galvanized by tragedy. When the steamship *Royal Charter* ran aground in Wales in 1859, crowds watched the wreck from the cliffs above, standing "in the leaden morning, stricken with pity, leaning hard against the wind, their breath and vision often failing as the sleet and spray rushed at them," as Charles Dickens described the scene. Only forty-one of the nearly five hundred passengers survived. Robert Fitzroy—who had been captain of Charles Darwin's ship, *Beagle*—sprang into action. Collecting every meteorological observation on which he could get his hands, he and his colleagues at the Board of Trade (to which he was meteorological statist, the contemporary word for statistician) drew hourly pictures of the storm's passage across Britain, showing the changes in pressure and temperature. Fitzroy called this new kind of map a "synoptic chart," and he hoped that the multiple observations shown would give greater insight into the weather. They predicted a technology that wouldn't come for another century. Much more expansive than "bird's eye," the synoptic charts were "as if an eye in space looked down on the whole North Atlantic at one time," Fitzroy wrote. Only twenty years after Ruskin first imagined his "vast machine," and only fifteen years after the telegraph was put into practical use, it was being used to save lives—not only locally, but across entire regions. The weather patterns the synoptic charts revealed were a boon in protecting the burgeoning steamship traffic of the Victorian era. Fitzroy's newly formed Meteorological Office soon had fifteen telegraph stations sending observations to London at eight o'clock each morning, and London would send back "forecasts." The first incarnation of the weather machine was up and running, as rudimentary but practical as the earliest locomotives.

Immediately, governments and meteorologists wanted more, and they set to work constructing as vast a machine as they could muster. They needed more observations, from more places, collected and shared in an organized fashion. Their project was both technical and political. Standardization was the rage, driven by the industrial revolution and the rise of international trade. In 1864, the International Geodetic Association sought to determine the size and shape of the earth. In 1874, the Universal Postal Union was founded. In 1875, the Paris Metre Convention defined the meter as a standard worldwide unit of measurement. Against this backdrop, the first congress of what became the International Meteorological Organization met in Vienna in 1873. Thirty-two representatives of twenty governments attended. They were primarily scientists and directors of weather bureaus, many of which had only just been established. Their fundamental project was to begin the international exchange of weather observations, with "observatories on islands and at distant points of the Earth's surface," in the words of Christophorus Buys Ballot, the founder of the Royal Netherlands Meteorological Institute and the first director of the International Meteorological Organization.

The diplomatic challenge was obvious from the start. If each nation was going to build its own weather observing system, and if those systems were going to be joined together into a system of systems, then they needed standards, protocols and rules. The delegates agreed that there should be two observation stations per each quadrangle bounded by ten degrees of latitude and longitude. But beyond that, everything was up for discussion. *What is the best form, size and mode of exposure of rain gauges? At what hour of the day should rainfall be measured? Can uniform times of observations be introduced? In what way should the proportion of cloud in the sky be estimated and indicated?* Many of the attendees at the first congress became proponents of Esperanto, the universal language for everything, which was no coincidence: They wanted a universal language of weather.

But they didn't yet have much to say. There were observations to exchange, but there was only so much meteorologists could do with them. The effort of constructing an interconnected and standardized network only highlighted how little they knew of how storms actually evolved. At best, they were reduced to a system of pattern matching. In a typical meteorological office of the time, "observing stations telegraph the elements of present weather," the English mathematician Lewis Fry Richardson later described. "These particulars are set in their places upon a large-scale map." The forecasters flipped through previous maps that resembled the current one, then moved forward in time from there, on the "supposition," Richardson noted, "that what the atmosphere did then, it will do again now. . . . The past history of the atmosphere is used, so to speak, as a full-scale working model of its present self." The limits of this method were plain. "It would be safe to say that a particular disposition of stars, planets and satellites never occurs twice," Richardson wrote. "Why then should we expect a present weather map to be exactly represented in a catalogue of past weather?"

By 1895, Cleveland Abbe, the architect of the US Weather Bureau and one of the most famous meteorologists in America, was fed up with these limitations. "Meteorology has been handsomely supported for a century by all governments and scientific organizations," Abbe wrote, in the very first issue of the journal *Science*. "Meteorology has received enthusiastic support by our own and all other nations. We are now doing about all that can be done by the mere utilization of the telegraph and weather map and the cautious application of general average rules, but we are still powerless in the presence of any unusual movement of the atmosphere." A vast *observation* machine was not enough. What meteorology needed was a new system of understanding— a theory. "Meteorologists can never be satisfied until they have a deeper insight into the mechanics of the atmosphere," Abbe continued. "Something more is needed than the most perfect organization for observing, reporting and publishing the latest news from the atmosphere. It is not enough to know what the conditions have been and are, but we must know what they will be, and *why so.*" Abbe concluded with a call to arms: "Further progress in meteorology demands a laboratory and the consecration of the physicist and the mathematician to this science." He had thrown up a flare—and it fell to a Norwegian working in Stockholm to rush over and help.

**In the most famous portrait of Vilhelm Bjerknes, he stands beneath a black um-**brella on the iconic docks in Bergen, Norway. His face glows in the sun, just then emerging from behind heavy clouds. In photographs his hair rises wildly to a halo; his chin is dramatically cleft, his eyes light. But in the painting, he appears mild mannered and content—a gentleman ahead of the weather, smug and satisfied with his capabilities of prediction. It's a good portrait for a meteorologist but maybe a little off-key for Bjerknes, whose contributions to meteorology were not empirical but theoretical. It was Bjerknes who first proposed the idea of calculating the weather—and then, despite massive technological limitations, worked out how to do it.

Vilhelm's father, Carl Anton Bjerknes, instilled in him both the math and the ambition. In 1881, when the younger Bjerknes was nineteen, the two of them traveled together to Paris, for the International Electric Exposition. The Palais de l'Industrie on the Champs-Elysées was filled with technological wonders. An electric tram—the first anyone had ever seen—ran through the giant nave of the building. Thomas Edison brought a twenty-ton generator from New York, nicknamed "Jumbo," which he used to power 1,200 lamps. Alexander Graham Bell's telephone was set up in a soundproof room, where visitors could listen to a live performance transmitted from the opera house across town: a "great blaze of splendor produced upon their minds." A "telemeteograph" automatically printed the current weather in Brussels every ten minutes. Surveying the technological wonders on view at the exposition, a writer for *The Electrician* magazine described one gadget on display in terms that sound like the narrative of a Victorian-era Apple video: "An absent love will be able to whisper sweet nothings in the ear of his betrothed, and watch the bewitching expressions of her face the while, though leagues of land and sea divide their sympathetic persons." The possibilities were limitless.

Carl Anton Bjerknes was a mathematician, and he had brought the teenage Vilhelm along to help demonstrate what he called "hydrodynamic analogies." If all around them Paris glittered with the visible expressions of electricity, their display inside the small Norwegian booth was devoted to its invisible behaviors. With his sleeves rolled up like those of a magician, Vilhelm put on a show meant to reveal the similarities between fluid dynamics and electromagnetism. One contraption—designed by the father and built by the son—had "two oscillating spheres mounted at the extremities of an arm," like a barbell, as *Popular Science* described it. Another had a sphere suspended in a tank of water and attached to a rod with a brush at the top, "arranged in such a manner as to paint, on the glass plate above, the line of every vibration of the fluid important enough to move it." Electricity was about to fundamentally change the world, illuminating cities and living rooms, and here was a vivid display of its otherwise invisible force. The crowd pushed in around Vilhelm. "I can hardly dry the apparatus with a cloth before people come back to see more," he said. It had nothing to do with the weather—not yet, at least. But the taste of acclaim fueled Vilhelm's ambition. His scientific work could be useful, if not famous. The elder Bjerknes was awarded a diplôme d'honneur for the display, the lone Norwegian on a list that included Edison and Graham Bell.

But back home in Norway, the Parisian lights and crowds behind them, father and son stagnated. Carl Anton was seized with self-doubt and couldn't write up their experiments for publication. Vilhelm picked up the work out of loyalty to his father, and some sense of his own professional opportunity, but to little effect. He returned to Paris to study with the mathematician Henri Poincaré, and then traveled on to Bonn, to work with Heinrich Hertz, the namesake of the unit of measurement known as the *gigahertz*. Bjerknes was looking for camaraderie and collective progress, but it eluded him. "Instead of spending his evenings at a physics kneip, drinking beer and discussing science, as he had imagined, [Vilhelm] found himself, with the lack of colleagues and the challenge entrusted to him, spending long hours alone," wrote his biographer, Robert Marc Friedman. When Vilhelm did finally finish his father's manuscript, he tried to play hardball with a publisher, demanding an advance, but they offered only "exposure" instead. For a moment in the late 1880s, Vilhelm resigned himself to this fate: a professional peak at nineteen in the illuminated exhibition halls of Paris, and then a life of obscurity, never again in quite the right place at quite the right time. Until, quite suddenly, he was thrust into the scientific spotlight with work that would prove, in time, nearly as important as Edison's prize-winning bulbs.

What made Bjerknes useful was a lost balloon. In the summer of 1897, the Swedish explorer S. A. Andrée set off in a hot-air balloon from Dane's Island, above the Arctic Circle in Norway's Svalbard Archipelago, on an audacious expedition to float to the North Pole and beyond, perhaps to Alaska. His big balloon—custom-made in Paris with funding from Alfred Nobel and christened *Eagle*—carried a crew of three humans and thirty-six pigeons, who had parchment cylinders attached to their tail feathers. Four days after Andrée's departure, one landed in the rigging of a sealing ship, but from then on the only news of *Eagle* was rumor and fiction. Forty thousand people had gathered at the train station to see Andrée off, and his disappearance created a sensation like that of Malaysia Flight 370 over a century later. Among the most consumed was Nils Ekholm, an expert on Arctic meteorology who had dropped out of the ill-fated balloon expedition, only to be haunted by the loss of his colleagues. Ekholm was struggling to answer the obvious question: Clearly, the fate of the balloon had depended on the winds, but what had the winds depended on? He had weather observations from the days surrounding the expedition's departure, but they were all at the surface of the earth. As to what was going on in the sky—the "upper air," as meteorologists call it—he had no data and no theory, limiting even any speculation about where *Eagle* might have ended up. Ekholm's view of the weather was frustratingly two-dimensional.

Bjerknes had been working on the third dimension, without entirely realizing it. Since 1893, he had been teaching at the *högskola* in Stockholm, the city's new (and less prestigious) university. The focus of his work had shifted: Rather than electricity, he was now exploring practical applications of classical—as opposed to theoretical—physics. In particular, Bjerknes was interested in the idea of *circulation*, which describes how forces act around a curve. It was well understood for ideal fluids, in which pressure and density are constant, but the atmosphere is not an ideal fluid.

It has areas of different pressures and densities, which work against each other, creating movement. This was empirically obvious to weather watchers—storms spun up and died off—but beyond the limits of physicists to explain mathematically.

Bjerknes developed a precocious hypothesis. He posited that when pressure and density are unequal, the unequal parts will torque at each other until they are equal, like magnets snapping into alignment. Bjerknes's "circulation theorem" could determine the direction and intensity of that circulation—in theory at least. The specifics of what that meant in the atmosphere (and for the weather) weren't yet clear; this was a long way from any kind of weather forecast. But he'd worked out a theory of the atmosphere, or at least a small first piece of it.

Bjerknes presented his work in a lecture at Stockholm's Physics Society, speculating a little on its possible applications for meteorology. Ekholm, sitting in the audience, wondered if the circulation theorem could be useful in determining the fate of *Eagle*. Bjerknes was a meteorology neophyte, and they put their heads together, syncing up Bjerknes's physics with Ekholm's understanding of the atmosphere. They couldn't figure out where *Eagle* went (its wreckage wasn't found for thirty-three years), but they realized they were talking about a new way to look at the sky, using physics and math, rather than analogy and intuition. Bjerknes could only guess about how the winds behaved, but his math made it a particularly useful kind of guess: a hypothesis that could be proven or disproven with observations.

When Cleveland Abbe, the esteemed American meteorologist, heard about this, he helped Bjerknes get his hands on an unusual data set collected by kite during a storm at the Blue

Hill Observatory near Boston. Bjerknes and his colleague J. W. Sandström assembled the data into a three-dimensional picture of the atmosphere that day. When they plugged these observations into the circulation theorem, they matched. The atmosphere behaved as the theorem said it should. Bjerknes knew he was on to something, and he began refining the work. Abbe sent more observations, and again the results were promising. While the meteorologists quibbled over the details of what the theorem said about the formation of the storm—and what it might say about future storms—Bjerknes was plenty pleased by the broader point: that atmospheric phenomena could be described with mechanics. Abbe was thrilled too. Here was a glimpse of the meteorological "theory" he had wished for: a physics of the weather.

Bjerknes knew exactly what he had. In a letter to Fridtjof Nansen, an Arctic explorer and one of the most famous Norwegians of the day, he articulated the bounds of his project: "I want to solve the problem of predicting the future states of the atmosphere and ocean," Bjerknes wrote. "I had previously closed my eyes to the fact that this was actually my goal, I must confess, partially for fear of the problem's enormity and of wanting too much." He saw—correctly—that with more theorems and more observations, meteorology could become a modern science: verifiable, repeatable, mathematical.

But there were two major speed bumps that slow down meteorologists even today. First, Bjerknes needed a better look at the present state of the atmosphere—he needed more observations. Second, Bjerknes needed to know how that state would change—of which the circulation theory described only one small part. Knowing-the-atmosphere Bjerknes called "the main task of *observational* meteorology. How-it-would-change became "the first task of *theoretical* meteorology." The challenge of the observational task was straightforward in concept, if difficult in execution. Meteorologists needed "simultaneous observations of all parts of the atmosphere, at the earth's surface and aloft, over land and over sea," Bjerknes explained. He needed the "vast machine" warmed up and ready to go.

The challenge of the *theoretical* task was less obvious but, ironically, more feasible for Bjerknes. Drawing on the work of his predecessors—giants like Isaac Newton, Leonhard Euler, Claude-Louis Navier, and Pierre-Simon Laplace—Bjerknes whittled the physics of the atmosphere down to seven equations, which required observations consisting of seven variables: density, pressure, temperature, humidity and wind velocity (as a vector, so it counted as three). The equations were like brushes that sketched the different ways in which air can move around. Altogether they painted a picture of the atmosphere, in its full dynamism. From a snapshot of a single moment, they could extrapolate the atmosphere's future state. Just as we can guess at the speed of a horse from a still photograph of its gait, Bjerknes could use these equations to calculate the weather forward in time. "Based on the observations made, the first task of theoretical meteorology will then be to derive the clearest possible picture of the physical and dynamical state of the atmosphere at the time of the observations," he explained. With that picture of the "directly observable quantities" of the atmosphere in hand, he could "calculate as comprehensively as possible all accessible data on the non-observable ones." Meteorologists needed to determine the "initial state" of the atmosphere through observations in order to determine the "evolution from one state to another" through calculations.

Bjerknes's equations weren't perfect, and solving them wouldn't itself create a weather forecast in any way we'd recognize today. But they were descriptive enough to act as hypotheses that could be proven or disproven with further observations, and they became the basis for what became known as the "primitive equations," which are still in use today. Despite the fact that the work wasn't useful for forecasting the weather, Bjerknes had done something remarkable: He showed how weather forecasts could be science experiments, easily repeatable every day. If you could solve for tomorrow's weather, you could prove yourself right when tomorrow's weather actually came.

But he couldn't solve for tomorrow's weather. There was the persnickety fact that Bjerknes's equations were functionally unsolvable. Six out of seven were partial differential equations that as Bjerknes himself admitted, "far exceed[ed] the means of today's mathematical analysis." Worse, they were interlinked, so partial solutions were useless. The wind depended on the temperature and pressure, and the temperature and pressure equations depended on the wind (among all the other variables). Finally, the equations didn't even say much about "sensible" weather—i.e., whether it was going to rain or snow. They merely indicated pressure and temperature at given locations, sometimes high up in the air. But the basic idea of calculating the weather using the principles of physics was unimpeachable.

In 1904, Bjerknes published what became meteorology's most

famous paper, "The problem of weather prediction, considered from the viewpoints of mechanics and physics," in the German journal *Meteorologische Zeitschrift*. It was well received, even if its practical applications were limited. Bjerknes still needed far more observations. Nations had made progress developing networks of surface weather stations, but taking measurements at higher altitudes remained rare and technically difficult. At the time of publication, the Wright Brothers had flown only a few times at Kill Devil Hills in North Carolina; within a couple of years they would be demonstrating their constantly improving flying machine across Europe. By 1910, commercial Zeppelins were crossing the continent, serving as a platform for additional upper-air observations, and demanding—for safety reasons—the collection of even more. Bjerknes used his growing notoriety to advocate for even more upper-air measurements, with burgeoning "aerological societies" formed to collect them.

But it still was nowhere near enough data. It was impossible to collect the necessary number of real observations of the atmosphere with which to even begin the calculations. By 1913, Bjerknes had to be honest about the successes and challenges of his method. "Now that complete observations from an extensive portion of the free air are being published in a regular series, a mighty problem looms before us and we can no longer disregard it," he blustered in a speech in Leipzig, where he had a professorship. "We must apply the equations of theoretical physics not to ideal cases only, but to the actual existing atmospheric conditions as they are revealed by modern observations." Still burning with ambition, he bemoaned the fact that his adopted field had fallen behind. "The problem of accurate pre-calculation that was solved for astronomy centuries ago must now be attacked in all earnest for meteorology." Why shouldn't the storms be as predictable as the heavens?

But once again—and always, really, with the weather—that was easier said than done. This new theory of weather forecasting exceeded practical capability. "What satisfaction is there in being able to calculate tomorrow's weather if it takes us a year to do it?" Bjerknes lamented.

But what if calculating the weather could take only a day? That, at least, would be a start.

# 2

# The Forecast Factories

In September 1913, Bjerknes received a letter from Napier Shaw, the director of the British Meteorological Council. Shaw had recently appointed a mathematician to one of the United Kingdom's more remote observatories, a place called Eskdalemuir, in Scotland. Shaw suspected that his work might interest Bjerknes. "He presented me the other day with what he called a dream of a palace at the Hague," Shaw wrote, a little incredulously. This "palace" would be like a concert hall, capable of holding five hundred people. Standing in a box in the center would be a conductor, who would read out weather observations. In the galleries all around him, all five hundred people, pencils poised, would calculate that weather forward in time—each human computer responsible for a specific section of earth. Shaw thought Bjerknes might like the idea. "I explained to him that you had already set out upon the programme and recommended him to digest what you had already done."

The imagination behind this practical—or at least semi-practical—idea was a lively one. Lewis Fry Richardson was born in England in October 1881—the same month the teenage Bjerknes was in Paris, conducting his demonstrations at the International Electric Exposition. Richardson had what his biographers sometimes describe as an "unorthodox intelligence." He loved electricity and machines, collected insects and kept a natural history diary, including observations of the weather. He liked to meditate, leading himself into what he called "intentionally guided dreaming," half awake and half asleep, "the 'almost' condition that is most advantageous for creative thinking."

It carried him to unexpected ideas. When *Titanic* sank, Richardson imagined a system to detect icebergs in the dark, with a whistle to make a noise and an open umbrella to catch any reflected sound—sonar, basically. Bored at work one day at a lightbulb company, he sketched his steampunk fantasy of the manager's office, with periscopes, a revolving desk and a visitors' compartment that tilted up when a floor pedal was pressed, ejecting its occupant. His real employer at the time was the Sunbeam Lamp Company; this imaginary counterpart, the Moonbeam Lamp Company, believed in "combining pleasure with duty," as Richardson noted wistfully in his notebook. A second pedal raised the chair through an opening in the ceiling to bring the "weary manager" to a rooftop garden—an idea that has found life in many a tech-company headquarters.

But it was drainage ditches that led Richardson toward the weather. While working for a company that harvested peat for fuel, he was asked to plan the best layout for their bogs. Rather than sketch a grid and leave it at that, he worked up a mathematical formula that considered the porosity of the peat and the flow of water after a rain. The formula was unsolvable—or, rather, it consisted of a series of interlocking differential equations that would take years to solve. Undeterred, Richardson instead worked out a graphical method for determining the ideal approximate locations of the ditches, by plotting his equations on a graph and looking for their intersection. Then he worked the problem backward again, using this "good-enough" answer provided by his graphical method to come up with an arithmetical procedure that gave him the same solution, as if measuring his angles with a protractor. The exercise satisfied him.

After National Peat's director embezzled funds and ran off to France, Richardson moved on to the Metrology—metal, not meteorology—Department of the National Physical Laboratory, bringing his math with him. He used it to calculate the stresses on masonry dams, with the actual calculations divided among a handful of "computers," meaning boys. The quickest of them, he reported, could do two thousand operations a week, which Richardson paid for by the penny, docking them for mistakes.

Having built a mathematical hammer, Richardson went looking for more nails. When he heard from Napier Shaw about Bjerknes's work, he began adapting the latter's equations to be solvable using his own methods. Then he went looking for an actual example to test them out on: real observations from a real day that he could use to create a real weather forecast, albeit retroactively. Bjerknes himself had just the right data set. He had recently published a series of detailed charts of conditions high in the atmosphere, made from observations collected as part of the "international aerological days" that he had helped organize. In May 1910, weather observatories across Europe had launched over a hundred and fifty balloons and thirty-five kites over a three-day period coinciding with the passage of Halley's Comet, which was expected to have an impact on atmospheric conditions. (It didn't.) The charts were organized into a bound book, as big as a broadsheet newspaper. Fourteen sheets corresponded to the conditions at specific altitudes, from sea level to the troposphere, from seventeen different locations across Europe—from Bergen in Norway, to Tenerife in the Canary Islands, to Pyrton Hill in Oxfordshire. It was an unprecedented snapshot of the atmosphere for Richardson to plug into his calculations, working toward a draft of what would become a book, which he at first called *Weather Prediction by Arithmetic Finite Differences.*

But World War I was on, and by May 1916 Richardson, who was a Quaker, could no longer avoid it. He was thirty-five years old when he arrived at the Western Front, with the 13th Section Sanitaire Anglaise of the Friends' Ambulance Unit. He wore a long beard, and his fellow ambulance drivers called him "Prof"—short for either "prophet" or "professor." In the daytime Richardson ferried the injured. In the evenings, he took out Bjerknes's bound observations and a twenty-five centimeter slide rule and worked on his calculations. His office "was a heap of hay in a cold rest billet." During the Third Battle of Champagne, in April 1917, he sent the manuscript to the rear for safekeeping, but it was lost, only "to be re-discovered some months later under a heap of coal," he reported. The unit's maintenance crew lauded Richardson as "a careful and conscientious driver [who] managed to avoid careless driving through shell-holes." But Richardson himself had a harsher assessment of his skills: "I was a bad motor-driver because at times I saw my dream instead of the traffic."

*Richardson's dream*. The phrase has become famous in meteorology, probably because his project is almost impossible to characterize in more practical terms. Richardson's goal was a six-hour weather forecast for May 20, 1910. "It took me the best part of six weeks to draw up the computing forms and to work out the new distribution in two vertical columns for the first time," he reported. "With practice the work of an average computer might go perhaps ten times faster." That was optimistic. When Richardson first conceived of his forecasting "palace in the Hague," he imagined his method could be accomplished by five hundred human computers. By the time Richardson published his book, in 1922, he had determined that an operational numerical forecasting office for the globe would require a computing staff of 64,000. The number made him sheepish, understandably. "The scheme is complicated because the atmosphere is complicated," he wrote.

But that didn't prevent him from capping off his plan with a wildly grand vision of what a working forecast factory might look like. "After so much hard reasoning, may one play with a fantasy?" he inquired, politely, in the final pages of *Weather Prediction by Numerical Process*. "Imagine a large hall like a theatre," he began, echoing the notion he had first presented a decade before. But if earlier Richardson imagined a mere concert hall, it had now grown to something more like a stadium, with walls reaching up into a dome, painted with a map of the globe, with England "in the gallery" and Antarctica "in the pit." Each human computer works on the equations for the weather in the part of the world in which they sit. Data is exchanged among them with "night signs," while their pace of calculation is dictated by an assistant standing on a high pillar who shines on them "a beam of rosy light" or "a beam of blue light," depending on whether they are running ahead or behind. "In this respect he is like the conductor of an orchestra in which the instruments are slide-rules and calculating machines," Richardson explained. To imagine a way to deal with the simultaneity of the earth required imagining an edifice in which to do it. What Richardson dreamed was what the contemporary mathematician David Gelernter calls a "mirror world": a database in space that represents an equivalent space in reality. Richardson's forecast factory was a kind of anticipatory memory palace—not a place to store memories of the past but a building that calculated the atmosphere of the future. Another way of describing it would be as "a model"—and indeed, its architecture anticipated the design of parallel computer processing, with many chips working side by side, that powers today's weather models.

But the fact that the sky over Europe on May 20, 1910, bore little resemblance to Richardson's prediction of it was a failure that haunted efforts of calculating the weather for decades. (The 64,000-person head count didn't help much either.) But even though his forecast was twelve years late (and wrong), Richardson was undeterred from the bigger idea, rightfully so. "Perhaps some day in the dim future it will be possible to advance the computations faster than the weather advances and at a cost less than the saving to mankind due to the information gained," Richardson mused. "But that is a dream."

What astonishes me is how thoroughly the forecast factory

anticipates the global view at the heart of the weather machine. Even amid the ashes of the Great War, Richardson could see the actual globalism that would define the system as we know it today, both politically and technologically.

**Bjerknes knew better than to try these calculations, but he wanted another crack** at the weather forecast. He spent the first part of the war in Leipzig, where he was put to work by the German war machine. He organized weather field observers and calculated wind speeds and direction for use by artillery units and for the dispersion of chemical weapons. Five of his assistants were sent to the Western Front and died there. He returned to Norway in 1917, fifty-five years old, still looking for a practical application for his ideas. The country was the perfect laboratory. The war had severely curtailed the international exchange of observations, all but eliminating western Norway's storm warning system and threatening its merchant fleets. There were food shortages, which put pressure on the success of the summer's wheat crop. Farmers needed accurate warnings, and so did the nascent airlines, just then coming to terms with the dangers of flight in a snowy country.

In 1918, with support from the Norwegian government, Bjerknes opened what would become known as the Vervarslinga på Vestlandet, or the Forecasting Division of Western Norway. He located the bureau for this "extended weather service" in the top floor of his own borrowed house, on one of the hills above the university in Bergen, a city on a fjord near the North Sea coast. There would be no attempt to calculate the weather, however. What would distinguish the Vervarslinga på Vestlandet was the number and precision of observations collected, as well as the new methods of their interpretation.

At the time of Bjerknes's return from the war, Norway had only nine weather observation stations connected by telegraph, three of which were on the West Coast. But the war had made a new set of resources available, in particular a network of U-boat lookouts, with wireless telegraph and spotting equipment that could also be used to measure the direction of the wind within a few degrees of accuracy. The Norwegian navy gave Bjerknes access to a boat for visiting islands and lighthouses up and down the coast, to train and encourage his new spotters. He immediately activated ten additional stations, and during the spring of 1918 he brought an astonishing forty more into use. By the start of that year's official forecasting season at the beginning of July, Bjerknes was ready to assemble a picture of the Norwegian atmosphere of unprecedented resolution—with ten times the density of stations compared to previous networks.

At eight o'clock each morning, the observers would send their observations by telephone and express telegram to the Vervarslinga på Vestlandet. A photograph from the era, likely staged by Bjerknes to promote the bureau, shows the scene. In the foreground, a woman sits with a telephone like an iron to her ear. She writes on a notepad, presumably recording the observation. Affixed to the wall are a pair of large charts, their edges curled up from use. An assistant—Bjerknes's son Jack—inspects a barometer placed on a specially built shelf, like an heirloom. At a long table, three young assistants sit with their pens and inks set before them. The one nearest the camera has his legs crossed under the desk, confident and relaxed. His hands are busy. At his feet is a wicker wastebasket, ready for the iterations of his analysis. The assistants would first write the raw data on the daily maps—the pressure and wind direction reading from each contributing station coloring in the outline map of Norway. Then they looked for patterns.

"What discoveries have we made today?" Bjerknes would ask each morning. By ten o'clock, the Vervarslinga would issue its forecast, good through the next day. By pinpointing shifts in wind and pressure, Bjerknes and his crew could identify "lines of convergence," as they called them, and the rainfall that normally accompanied them. Representing them, their maps swirled with big arcs, showing no regard for traditional political boundaries, land or sea. Adopting the martial language of the era, they referred to these as "fronts"—an idea they extended to describe "polar fronts," which could be thought of as the battle line between polar and tropical air masses. It was an insight that stuck, with "Bergen School" methods finding their way, over the next generation, deep into universities and weather bureaus across the United States and Europe. Bergen School methods were used in the forecast for the Allied invasion of Normandy, the accuracy of which was a key element of D-Day's surprise. But for all that, they were not, strictly speaking, theoretical.

Sverre Petterssen, a Norwegian meteorologist who arrived in Bergen in 1923—and ultimately worked on that D-Day forecast—found these new methods thrilling, though he was well aware of their limitations. As an undergraduate meteorology student in Oslo, he had been frustrated by the "old and sadly outmoded" textbooks, "full of tabulated data and dreary descriptions of individual phenomena, with hardly any reference to the laws of physics," as he wrote in his memoir. Petterssen found the techniques of the Bergen School, in contrast, fresh and scientific, but only to a point. The mapmaking in Bergen and beyond was merely making a "series of simple mathematical expressions for the velocity, acceleration and rate of development of weather front and storm centers, without asking *why* and *wherefore*," Petterssen lamented. "I had to be satisfied with just, *it is so*."

Bjerknes was less recalcitrant and happier to champion his practical forecasting success—despite its contradicting his earlier insistence on calculating the weather. "During fifty years meteorologists all over the world had looked at weather maps without discovering their most important features," he bragged. "I only gave the right kind of maps to the right young men, and they soon discovered the wrinkles in the face of Weather." It's an evocative summation, but it's strange that Bjerknes, who evangelized for a mathematics of the weather, retreated in the end to more analytical methods. This was a necessary scientific compromise that proved—at the time—eminently practical. The historian Frederik Nebeker is sharp in his assessment of this shift: "It is ironic that the man who became known as the advocate of calculating the weather, and as the advocate of meteorology based on the laws of physics, was also the man who initiated the development of a set of effective techniques that were neither algorithmic nor based on the laws of physics."

But Bjerknes's greatest contribution to meteorology is simple: He showed how the scientific method could be applied to weather forecasting. Each calculation of the weather could be a hypothesis, proven or disproven by the weather (when it actually came). His focus on the intensive collection of observations, and the use of further observations to verify his calculations, showed how the abstractions of his math and the vicissitudes of the weather could be linked.

Bjerknes saw that weather forecasting is the archetypal example of what scientists call a "prediction problem." They come in all forms, from trying to predict the transmission of a disease, the behavior of the flame of a Bunsen burner, or the fragmentary trajectory of an explosion. Each can be tracked to the scientific method, with its cycle of hypothesis and verification. But the weather is special in that its predictions need not be constrained to the present or the near future. A new way of calculating the weather can be tested on the entire history of weather data. If a weather prediction is wrong—which it always is a little—scientists can try again a different way, tweaking aspects of their equations like an optometrist does a phoropter. Bjerknes and Richardson had only the slimmest hoard of weather data with which to work—and certainly would have benefited from a supercomputer—but their faith in the potential of their ideas was well placed.

What they also made clear was that to know the atmosphere of the future required knowing the atmosphere of the present. To know what will be, we must know what is—everywhere, all at once. The vast machine had to exist—but the planet we live on was indeed vast. How could its sky be seen and measured?

## PART II

# OBSERVATION

# 3

# The Weather on Earth

There is no weather forecasting without weather observations, but there are no weather observations without infrastructure. There are instruments attached to the noses of airliners, in fenced clusters alongside highways, in quiet corners of schoolyards and the backyards of hobbyists. A friend in Brooklyn has one lashed to the railing of his roof deck, next to the grill. It looks like a big bath toy, with a bulbous white plastic body and a spinning black anemometer. Its controller sits on his desk downstairs: a gray box with an LCD screen, reporting the temperature, wind direction and speed. It's of a type that could be connected to the Internet, but he hadn't bothered. Someone else nearby, a stranger, had already connected their station to the website Weather Underground, and when he checked his own location its readings were almost always identical. What would be the point? How different really was the weather across the street? In the scheme of things, his observations were redundant, but that didn't make them any less worthwhile for him. They brought order to his patch of sky.

There are weather stations most everywhere on earth, but not all weather stations are created equal. In recent decades, the most important sites have been part of what's been known as the Regional Basic Synoptic Network, managed by the World Meteorological Organization, the United Nations agency responsible (among other duties) for coordinating weather observations. The Regional Basic Synoptic Network consists of around 4,400 surface stations around the world, mostly run by national weather services according to standards of quality and calibration. The precise number of active stations is always in flux, as is the bureaucratic structure of the program itself. But what persists is a hierarchy of weather stations, with some thousands around the world operating at a higher standard of equipment and attention.

You can often spot one of these important sites waiting for takeoff at an airport, visible as a clutch of equipment beside the runway. At LaGuardia Airport in New York, the weather station is installed on a small patch of grass, burnt to brown by jet exhaust, at the edge of taxiway DD. It has an odd-looking steel bucket, raised a few feet off the ground and surrounded by twinkling metal fins, like a medieval hula skirt; it measures rainfall. Beside it are two poles mounted with cylindrical sensors. Each is pointed at another small sensor held up a few feet away by a steel bracket, like a skinny robot staring into a makeup compact. One measures visibility, the other handles "precipitation identification"—or whether or not it's raining or snowing. A ten-meter-tall tower, painted white and pink (faded from red) has an ultrasonic wind sensor at its peak.

Often, weather stations contributing to the Regional Basic Synoptic Network have a human observer on duty. At LaGuardia, the precipitation detector can't distinguish rain from ice. A ceilometer measures cloud cover, but only directly above the airport. Even if the thickest fog bank were rolling in from the west, over Manhattan, the machine wouldn't register it until it arrived. To compensate for these limitations, and to back up the automated system, LaGuardia is one of 135 airports around the United States with a human observer—a person watching the weather and watching the machines that watch the weather. The dayshift observer at LaGuardia is Paul Sauer. Among other things, while the ceilometer only looks straight up, Sauer, who has a doctorate in philosophy, looks in all directions.

But that effort to observe and measure the weather is consumed by tensions, between the continuous atmosphere of the earth, and the political borders that carve up the land; between the international cooperation that manages the system and the autonomy of the government meteorological agencies that operate its individual parts. The Regional Basic Synoptic Network is one component of what's grandly known as the Global Observing System, which is itself part of what's even more grandly known as the World Weather Watch. It is almost always described with an infographic that pops up regularly at weather stations and meteorological offices. The graphic renders the earth in three unequal realms: a blue sky, a darker blue ocean and a green land. Arranged in each realm are pictograms of the observing system's parts, labeled all in caps: OCEAN DATA BUOYs, which float offshore and are notoriously expensive to maintain; AIRCRAFT, which depend on the participation of the airlines; UPPER-AIR STATIONs, where government weather services launch balloons twice a day; and AUTOMATIC STATIONs, which may be as simple as a sensor mounted to a traffic light. Connecting them are red lines ending in arrows, which all eventually lead to a human figure sitting at a computer terminal in front of a map of the round sphere of the earth. Beside this weather humanoid is an old-fashioned mainframe computer, with reel-to-reel storage tapes, ostensibly to represent the weather models. What it seems to show is a whole world full of thermometers and anemometers and satellites and buoys, all flashing their data in the same direction: toward a person in front of one computer, connected to another bigger computer, containing a numerical simulation of the atmosphere. It portrays the global observing system, and the models it feeds, as a single coherent thing. This basic anatomy seems simple enough. But it belies a profound tension between the specificity of each observation station—which samples the atmosphere in one place in one neighborhood; or one buoy in one harbor; or one airliner on one route—with the reality of the continuous swath of the atmosphere of the earth as a whole. In Ruskin's terms, it shows the space but not the point.

It is easy to say that the weather forecast depends on observations made all over the world. It is harder to bring into focus the individual stations, those tens of thousands of real places, sampling their little squares of atmosphere. To see them as generic equipment clusters, magically connected into a global network, is to misunderstand the often long, and always local, history of each place—what I think of as the inertia within the system. The lore of weather stations is deep and wide. When Hurricane Florence approached the North Carolina coast in 2018, for example, a decommissioned Coast Guard lighthouse thirty-four miles out to sea called Frying Pan Tower had a moment of celebrity as it broadcast the storm's arrival—a forgotten place suddenly remembered thanks to its unique view of the weather. In New York City, the National Weather Service office was for decades located at 30 Rockefeller Plaza—with its equipment on the roof, just above the famous red neon sign—as apt and important, if incongruous, a site for weather watching as any lighthouse. (It later moved to a less glamorous, if roomier, location on Long Island, on the grounds of the Brookhaven National Laboratory.) Then there are the wild and remote stations like Jan Mayen, a Norwegian volcanic island way out in the Arctic Ocean, north of Iceland and halfway to Greenland. It has a staff of eighteen people and two dogs and is nearly impossible to visit. (A military cargo plane flew in eleven times a year.) But when it comes to weather watching that region of the world has an allure I couldn't shake, in part thanks to Bjerknes.

When Vilhelm and Jack Bjerknes published the results of the Vervarslinga på Vestlandet's first summer of intense forecasting, they included a kind of atlas of their accomplishment: an outline map of Norway, showing the locations of the observation sites dotted across the country like a tightly patterned shirt. I could imagine the assistants at the house on the hill in Bergen working the phones and sketching their fronts. But I had more trouble picturing the other end of the line—the places represented on the map. Were they on farms or the roofs of schools? Above a harbormaster's office or beside a lighthouse? Bjerknes's map of weather stations was drawn not to illustrate the fluid atmosphere but its more fixed human mirror. It wasn't a weather map but an infrastructure map. These weather stations were crucial to Bjerknes's insights, but they were unspecific, elided away. I already saw how easy it was to take "the observations" for granted and assume they just kind of happened, all at once, an invisible system of worker bees and measuring tools. But what was each one really like? A vast global machine of weather observation had been erected over a hundred and fifty years. Bjerknes added more than his share. I thought that if his observers could come back—or if they never left—their stories would be essential to understanding how these observations have evolved over time, and what it actually means to "observe" the weather, rather than simply experiencing it, as we do every day?

Gabriel Kielland knows the geography of Norway's weather stations better than anyone. He works in Oslo at the Meteorological Institute, leading the department responsible for observational quality. Before my lunch with Anton Eliassen, I sat down with Kielland in his high-ceilinged office in their red brick building, built in 1939, at the edge of the city adjacent to the university campus. Outside was a cluster of observation equipment, installed on a lush green lawn like a sculpture in an office park. On his computer screen, he called up the current status of Norway's weather stations—a list of about four hundred, not all of which were his responsibility. An additional several hundred were alongside highways and belonged to the road authority. Others were on offshore oil platforms, which were steadily replacing the old lighthouses as observation stations. Everything working well was highlighted in green; anything with a recent problem that had been fixed was in yellow; the broken ones were in red. "So there is an issue right here," he said, looking down the list. The rain-collecting bucket just on the other side of the window needed to be emptied. Kielland giggled, unembarrassed.

When I shared Bjerknes's map with him, he gamely took a stab at identifying the locations of the observation stations. Many of the places he could recognize with certainty. "The first weather stations were the telegraph points," he explained. Norway's unusual geography meant that all but one of them were on the coast—often close to the lighthouses, which were mostly positioned in the approaches to these cities. "So they are quite strategic." But a handful of the dots on the map made Kielland suspicious. Some he suspected either didn't exist at all or were deliberately misplaced. "Apart from my accusation of fraud it is clearly a major achievement to collect this many wind observations," he said dryly. Observation sites had a habit of becoming fixed, as if in rock. Which stations were still there? Maybe there was a hundred-year-old weather station, still sending its observations back to Bergen, or on to Oslo, I asked. Perhaps one of the lighthouses?

"The lighthouses are all remote, but the best would be Utsira, which is between Bergen and Stavanger," Kielland said. "It's been there since the beginning, since the 1860s. That's a good place."

Utsira is a speck of an island the shape of a Rorschach blot ten miles out in the North Sea. It may only be a single point, but it was single points that I was after. It is through single points that we understand the whole. Each place has a story to tell, but only when they are all linked together do they tell the whole story: a picture of the atmosphere of the earth at that moment—

the necessary starting point for drawing a picture of the atmosphere at the next moment. In the database of the Regional Basic Synoptic Network, Utsira's "station identifier" was 1403. Its location is recorded as 59° 18' 23" north latitude, by 4° 52' 20" east longitude. That metadata matters—it's the structure that makes Utsira's observations useful to the world, allowing them to be plotted on the map or ingested by a supercomputer. But I prefer to think of Utsira in human terms: When we look at the forecast for tomorrow we often forget (if we ever knew) that it's someone's job to look at the sky there.

**The flight to Utsira was smooth and silent, without a hiccup or rattle from the air** outside. As the plane crossed the mountains, I couldn't stop observing those Norwegian clouds. They were fluffy and distinct, as if each one had been cast from a different mold. They looked like parading elephants: big, bulky things, evenly spaced, dark on the bottom and luminescent on the top. We glided between them into Haugesund, a small city midway up the North Sea coast. There was no jetway, but the air was springy and damp as I crossed the tarmac to the single-story beige terminal building, with a big circular window set in a bright blue square. Above it, in painted letters, was written in English: "Welcome to the Homeland of the Viking Kings."

I traveled to Utsira through Haugesund, which sits on the Smedasundet, a little strait that connects to a bigger strait, the Karmsundet, across whose protected waters Thor, god of weather, is said to have waded each morning. In more recent years, the

Karmsundet offered a shipping route that had made Haugesund prosperous. As I neared the dock for what everyone calls "the *Utsira* boat," the most recent evidence of the city's success came into view: a floating yellow power converter, twenty-five stories tall, looming over the harbor like a four-legged monster. Built in Dubai and towed to Norway, it was almost ready to be installed alongside a wind farm in the North Sea, where it would marshal electricity for a million homes in Germany. This was the latest incarnation of the "offshore" boom. In the nineteenth century it was herring; more recently it was oil; now it was wind too. Norway declared sovereignty over the North Sea's oil and gas deposits in the 1960s, with Utsira helping to stretch its territorial claims a bit farther out to sea. Since then, Norway's Oljefondet, or Oil Fund, has collected over a trillion dollars in tax and license fees, and the country peaked as the eighth largest oil exporter in the world.

The ferryboat *Utsira* reaped some of the benefits. It was a brand-new, tall, squat, tough-looking little ship, with a sweep of blue stretching from its improbably high deckline down to the water. Its big bow doors were open, and its vehicle bay was empty. No one was around. I climbed a steep flight of stairs to the passenger lounge, decorated with ficus trees and pretty black-and-white photographs of the island Utsira in Ikea frames. A television was tuned to *Hardcore Pawn*, an American reality show. On YouTube, I'd watched a terrifying video of a boat riding waves like roller coasters in the waters we were about to cross. But the sea was calm, lapping gently at the dock. We slid out between rocky shores lined with modernist townhouses and ancient-looking fishing shacks, while a bearded deckhand in Birkenstocks and rubber pants collected my eight-dollar fare. *Utsira* was an inexpensive time machine, carrying me back to the weather infrastructure of the nineteenth century.

**Utsira had a weather observer before Norway had a weather service.** In the beginning and still today, the island's observation point is a patch of grass between a pair of twin lighthouses, set on either side of a small saddle of land near Utsira's highest point. Known in Norwegian as Utsira Fyr, the lighthouses were built in 1844 to guide ships around the herring fishing grounds and on their way to Haugesund. Utsira was itself once an important port as well, with two government-built harbors, one on each side of the island, to shelter the ships regardless of the direction of the brutal winds. At first, that wind was measured by the lighthouse keeper, using the scale invented by the Royal Navy officer Francis Beaufort. The keeper would look out across the island and judge the movement of smoke and trees. If smoke rose vertically, that would be Force 0; if large branches swayed it was Force 6. When Henrik Mohn became the founding director of the Norwegian Meteorological Institute, in 1866, he added temperature measurements to the keeper's duties, with instructions to mail postcards with observations thrice daily. When the telegraph came to Utsira in 1869, the island became a crucial lookout for North Sea storms. For Britons, Utsira—or at least its name—retains a sentimental notoriety: "North Utsire" and "South Utsire" are two of the areas of sea chanted in the "Shipping Forecast," still broadcast on BBC's Radio 4, accompanied in the evening by a cheesy arrangement of "Sailing By."

Topography is everything on Utsira. Its lush green hillocks are scattered with granite boulders, like a giant miniature golf course. Its houses hold on tightly to the rocks, with flags like sails keeping them straight to the wind. The biggest and tidiest of them belong to "oil rig bosses," as an islander described them to me, who spend two weeks offshore and four weeks on, commuting by helicopter. But topography is nothing without geography and the economic and political advantages it brings. Utsira's importance as a place from which to observe the weather is as much political as meteorological. Utsira has never been merely a neutral post to observe the atmosphere but always a political pawn. The island is disconnected, as islands are, but also perpetually drawn into the orbit of broader powers, like the ribbon in a game of tug-of-war. For the most part, and mostly benignly, that influence comes from Oslo. Even today, Utsira is dependent on the federal government's largesse, which is considerable.

Utsira Fyr's modern weather instruments are installed at the edge of a courtyard at the base of the lighthouse's little hill. Temperature, precipitation, humidity, wind speed and wind direction are all transmitted automatically. But there's also a human involved, for backup and to make the kinds of nuanced observations the automated equipment can't. Utsira's part-time observer, employed by the Meteorological Institute, is one of sixty scattered across Norway, measuring the cloud cover and characterizing the precipitation. In other places, the weather observers tend to be farmers doing the work originally performed by their parents and grandparents—a human link of land, people and weather. For thirty years the weather in Utsira had been observed by Thorbjorn Rasmussen, who was also the island's mayor and its last lighthouse keeper. When he retired, Hans Van Kampen, a Dutchman, took over.

I met Van Kampen at the lighthouse just after lunch. He has a weathered red face and unruly, enthusiastic, red-gray hair. He and his wife first came to Utsira in 2006, fell in love with the place, and stayed to take over the island's pub. When the schoolhouse was replaced by a crisp glass-walled building set into the hillside, they bought the old building, moved in upstairs, and opened a new restaurant on the ground floor. Their customers are mostly day-trippers to the island, but only on calm days. No one wants to suffer an hour of seasickness for a nice lunch and a little bird-watching. "People are afraid of the wind here," he said. "The more wind, the less come." The local business is steadier; Van Kampen has the only espresso machine on the island. His customers know he works for the Meteorological Institute and often ask him for a forecast, without realizing his purview is the present and not the future. To satisfy them, he makes a habit of looking at the Meteorological Institute's public website, called Yr (which means "mist"), and repeating what he reads there. Six times a day, Van Kampen looks up at the sky, often while standing at his back door with a cigarette. His predecessors sent postcards; Van Kampen uses a series of dropdown menus on the Meteorological Institute's website to transmit his report to Kielland's department in Oslo. Oslo then sends it onward to the world—where it joins the streams of data flowing through the World Meteorological Organization's Global Telecommunication System—and eventually finds its way into the weather models and back to the forecasts on Yr. When the future becomes the present again, Van Kampen is there to observe it.

The other part of his duties consist of maintaining the automatic observation equipment at the lighthouse, and so, after a cup of coffee, Van Kampen put his wellies on ("We have a lot of sheep shit") so we could take a closer look. The station's location may not have changed in a hundred and fifty years, but the current setup was state-of-the-art. The instruments lived inside a box with a gable roof, like a dollhouse, raised improbably high off the ground. I climbed up a short wooden staircase set at its base, like the ramp to a chicken coop. Van Kampen is extremely tall, and this put us at eye level. He undid a rusted latch, and we peered inside together. Mounted on the box's left wall were the temperature and humidity probes. They looked like little white umbrellas set into a stand, their rubber cables looping upward. The humidity sensor had been calibrated in Kielland's basement laboratory in Oslo, using a machine that looked like an iron lung, and then shipped to the island. Opposite them were a pair of old mercury thermometers, crusty, stained and unused. Mounted outside the box there was a precipitation sensor, with a flat surface open to the sky, like a dinner plate. The telecommunications equipment that obviated the need for a human observer to send in the temperature and wind readings was bolted to the wall in a small room in the compound's garage: two steel boxes the size of microwaves, with cables snaking in. Just behind the old keeper's house, in a bog, was a sophisticated rain gauge, with metal fins that sheltered its bucket from the wind and swung around like wipers at a car wash. On the ridge next to the lighthouse was an ultrasonic wind direction and speed sensor, a high-tech replacement for the old-fashioned anemometers that had spun on Utsira since 1932.

It was time for Van Kampen's midday observation. We stood at the base of the thermometer enclosure, and he took two small books from a canvas satchel. The first was his Meteorological Institute observation guide, printed in color and spiral-bound. On page 19 was a list of numerical codes, 0–9, that described the amount of cloud cover—a protocol Henrik Mohn advocated for at the first congress of the International Meteorological Organization in 1873. The second book was Van Kampen's official journal, the paper trail of his electronic submissions. Van Kampen arched a bushy eyebrow to the sky. I matched his posture, trying to see what he saw. "Well, there's drizzle. Ah?" he asked. Yes, I concurred. It was drizzling. He scribbled. "Sky height? I'd say three, four hundred meters. Visibility? How long can we look? We can't see the end of the island, so maybe two kilometers? No more than that. Type of cloud? I don't know how to translate. *Mist clouds?* Just gray cloud cover. When those are the only clouds you can see, it's very easy. There's nothing up there. Or there is something up there, but you don't see it, so I don't have to report it!" Van Kampen laughed, mostly at himself, but I thought a little at me too.

It is tempting to romanticize Van Kampen's life on Utsira, and watching the sky and making espresso on the edge of the world indeed seems romantic, out of time and rooted in place. But it is also, by its very nature, quotidian. Van Kampen watches the weather in this one place so that the rest of us can know the weather in all the places. That's how weather observation stations have always worked. I wanted to ask him if this project felt somehow bigger than its parts, if it summoned some profound understanding of the vicissitudes of the weather and his place on earth, reaching back to his childhood conception of the world. But what I ended up asking him was if he *really knows* the weather?

"I know when it's cold and when it's wet," he replied, quashing my hopes of some Thoreauvian mysticism. Then he excused himself. He had to get back to the restaurant.

When I was alone again, I climbed up the little hill to the lighthouse. I could see the sea all around, but it wasn't like being on a ship. The ground was firm, made of grass and moss and granite, strewn with dung and tufts of white fleece. The famous Utsira wind roared around me. I was walking around the tower, eyes on the ocean, nose to the breeze, hands in my pockets, when I tripped on something. Inserted into the granite were four bolts, each about six inches tall—some detritus of the past century of weather observation at this spot. When I got up, my jeans damp from the moss, I had a sudden sense of shifted perspective: Here was the lever of Archimedes, the place where you stand firm to measure the air. I felt the friction between the wind and that anchor to the rock of the earth—felt acutely here, in this single spot, where the land juts out of the sea and the atmosphere flows by. It's what wind *is*, I realized.

The essence of a weather station is to stand firmly in place to measure the atmosphere rushing by. That contrast between the static and dynamic is a big part of what intrigued me about the weather and the ongoing human project to watch and predict it. Not many places are as fixed in time as Utsira, as unchanging. That makes it easier to grasp the daily rhythms of the weather and the necessary vigil of its observation.

Places like Utsira are a crucial part of the global system of weather observation, not least for their long and continuous history. But still, Utsira is just a speck on the map, and no number of specks could ever be enough. To accurately predict the weather, Bjerknes's equations needed more—a more expansive and continuous view. Utsira and the surface observation stations came out of the era of Fitzroy, in the nineteenth century, although even then he knew the view he wanted, even if he couldn't get it: "as if an eye in space looked down on the whole North Atlantic."

It would finally come in the wake of World War II.

# 4

# Looking Down

World War II was hard on Utsira. During times of conflict, its remoteness didn't remove it from the consideration of higher powers but forced it into play. Four hundred Nazi soldiers occupied the island, turning its central valley into a garrison and parade ground; its narrow lanes lined with wildflowers were littered instead with military tents and smoldering oil drums. The smaller of Utsira's two lighthouses became an anti-aircraft gun nest (today it's a cellphone tower), and the other went dark, its upper deck put to use as a lookout over this strategic section of sea. When I climbed up there, I could see the chip in its glittering Fresnel lens, manufactured in Paris in 1890, left by the butt of a clumsy Nazi's sidearm as he circled the darkened lantern room. Controlling Utsira had long-term benefits, because one never knew when it might be crucial as a marker on the map, an extra port in a storm, or an observation point of the sky and the sea. During the Cold War, the lighthouse keeper became a NATO lookout, watching for Soviet subs, ships and planes. A Geiger counter was still on display, a haunting memento of the stakes.

This is often the case with weather stations. In peace, observing the weather is utility work, like swabbing the decks and trimming the trees. But in war, observations become secrets, and weather forecasts are weaponized. The desperation for them has led to expanded networks and new technologies.

World War II marked the beginning of a transformation of weather observation from a collection of disparate points into a global system—made up of observatories on the ground, in the air and, soon enough, in space. But it happened piece by piece, driven by technological developments and military needs. The fighting in the North Atlantic stretched from Labrador and Greenland in the west, to Svalbard and Franz Josef Land in the Barents Sea, all the way east to Novaya Zemlya, which separates the Barents from the Kara, north of Siberia. Throughout, the Germans were at a distinct meteorological disadvantage. The Allies predominantly held the northerly and westerly positions, while storms tend to move from west to east and from north to south. Before the war, remote whaling stations in Greenland and Iceland would radio observations for the benefit of ships throughout the region. But just as the Civil War broke apart the Smithsonian's early observation network in the United States, World War II halted the exchange of weather data across the North Atlantic.

The Nazi weather service, known as the Wetterdienst, moved quickly to compensate, sailing observation ships up the North Sea and into the Arctic, with a meteorologist onboard to launch balloons. When the Allies began to sink the unarmed ships, the Wetterdienst turned to new technological solutions. The Siemens-Schuckertwerke corporation—predecessor to today's Siemens conglomerate—developed an automatic weather observatory, code-named *Kröte*, or "toad," with nickel-cadmium batteries and a powerful radio to transmit readings. The earliest versions were small enough to be delivered to remote locations by airplane, but keeping them hidden and operating was a challenge. The first toad, deployed on the Norwegian island of Spitsbergen in 1942, was quickly found and dismantled. The second, on Bear Island, had its antenna destroyed by bears. With more than two hundred U-boats patrolling the North Atlantic, working to maintain the blockade of England, the Germans' need for weather observations became desperate. By the fall of 1943, Siemens had developed a new version of the toad with a ten-meter antenna, powerful enough to broadcast encoded observations all the way from the North American coast to receiving stations in Europe, but small enough to fit inside a submarine's torpedo tubes.

U-537, sent to install it, sailed from its concrete pen in Bergen, Norway, on the night of September 30, 1943. Its destination was a spot near the present-day border between Labrador and Quebec—a location the captain hoped would be far enough south to be free of ice and far enough north to be free of locals. A photograph, unearthed from the military archives in the 1970s, captures the scene upon their arrival in North America: seven sailors in black knit caps standing around two rubber dinghies laid askew on the U-boat's deck. Working in the autumn fog, they lugged ten gray canisters, each the size of a large bucket and weighing two hundred pounds, to the top of a nearby hill. When they had finished assembling the system, they hand-painted "Canadian Meteor Service" on the canisters and littered the site with

American cigarette packs. Even today, with the familiarity of satellite communications, solar panels and small sensors everywhere, it seems an audacious idea: a clandestine intercontinental automatic weather station, a Wetter-Funkgerät Land. Given the designation WFL-26, the remote weather station broadcast for less than a month before its transmissions were mysteriously jammed.

And then it disappeared for forty years. A US Navy team missed it in 1952 while scouting the area for places to install the massive radars of the DEW Line, the Distant Early Warning network, built to watch for Soviet long-range bombers. A Canadian geomorphologist stumbled upon it in 1977 but assumed it was, in fact, what it said to be on the can: an automated Canadian Weather Bureau station. It was only after a retired Siemens employee and historian, Franz Selinger, noticed the unusual landscape in the photographs that accompanied the logbook of U-537 that anyone went looking for it at all—evidence of the only known Nazi incursion on North American soil. Sailing on a Canadian icebreaker, he and a Canadian military historian finally found it in 1981, systematically vandalized, its connections cut and contents strewn across the rocky ridge. Today, "Weather Station Kurt"—as it became known, named after its government minder, Kurt Sommermeyer—is on display at the Canadian War Museum in Ottawa, looking gray and ugly, like so many weapons. (U-537, for its part, is at the bottom of the Pacific, sunk in 1945 by the submarine USS *Flounder*.)

It's a wild story—an act of meteorological desperation and technological bravado begging for Hollywood. But it marks a pivot point in the history of weather observation. For the first century of telegraph-based observation networks, meteorologists were preoccupied with expanding their reach, occupying lighthouses, ships and airfields. The war cut the map in half. But it enabled new technological advances that raised the possibility of a newly expansive view of the weather. For the hundred years before Weather Station Kurt, the telegraph allowed for news of the weather to be transmitted faster than the weather itself. But someone had to be there. (Often, somebody still does, as on Utsira.) Kurt was the proto-example of a new kind of station, capable of working on its own. Soon, these observatories would not only be in remote corners of the Atlantic but high up above the earth.

**Another Nazi technology opened up new possibilities for observing from the sky.** In the final months of the war, Wernher von Braun, the German rocket engineer, made a terrifying and deadly technological leap, successfully launching the first guided missile, known as the V-2, or "vengeance weapon." It was horribly inaccurate, flying far wide of its intended targets, but still killed nine thousand people in London, Antwerp and Liege. After the war, the United States and the Soviet Union famously scrambled to collect the remaining rockets and the scientists who designed them, and von Braun was brought to the United States. In the earliest days of the Cold War, their priority was adapting the V-2 for continued military use—and indeed, its design would serve as the basis for both Soviet and American rockets eventually capable of carrying nuclear warheads, and astronauts, to space. But first they would be used to observe the weather.

In October 1946, technicians at the White Sands Proving Ground in Nevada installed a camera in the nose cone of a captured V-2 and launched it straight up into the sky. Within thirty seconds, the rocket disappeared from view. But then it began looking back, its 35 mm camera snapping photographs every second and a half, up to an altitude of eighty-three miles, before crashing down to the desert. A search plane found the wreckage and recovered the film, protected in a cylindrical steel cassette the diameter of a dinner plate. "A truly dramatic spectacle unfolded when the film was developed," the camera's designer, Clyde T. Holliday, recalled. "On these photographs we saw what a passenger on a V-2 would see if he could stay alive on the zooming ride up to that height and back again, and how our earth would look to visitors from another planet coming in on a space ship." It was a view that had before been only imagined, but its practical benefits were unmistakable. This first visit of a camera to the margins of space yielded photographs of a quarter of the United States, an area of nearly a million square miles. The curvature of the earth was visible, along with bands of clouds stretching hundreds of miles in rows like streets.

Meteorologists were immediately enthralled by the possibilities. The director of the Weather Bureau, Francis Reichelderfer, wanted reprints of the images from the camera's designers at the Applied Physics Lab at Johns Hopkins University to be shared with every weather office in the country, "so our forecasters can obtain a glimpse into what may well turn into a potent weather forecast tool in the future." They couldn't yet imagine actual orbiting satellites, so instead Holliday, the camera's designer, began thinking through how this single experimental camera strapped to a rocket could be extended into a full weather-observing *system*. "If guided missiles carrying cameras could be sent out criss-cross over the entire continent of North America every day, photographing in a few hours all the cloud banks, storm fronts, and overcasts, weather forecasts could be made more accurately than now," he wrote, somewhat elaborately. There was no sense that this would happen. At best, it would be a stopgap measure, useful only until rockets were powerful enough to enter orbit.

The prospect was tantalizing. Working out what an actual weather-watching spacecraft might be fell to the RAND Corporation (a contraction of "research and development"), famous for having its hand in all kinds of complex technical systems of the era, from nuclear war planning to the nascent Internet. "Can enough be seen from such altitude to enable an intelligent, usable weather (cloud) observation to be made, and what can be determined from these observations?" the authors of the top-secret 1951 report "Weather Reconnaissance from a Satellite Vehicle" wondered. Their concern was that the satellites would only look, not measure; they would take pictures but provide no numbers, none of the quantitative measurements upon which meteorologists now relied. They would undoubtedly steer weather forecasting in a particular direction—back toward empirical methods—while doing nothing to support efforts to calculate the weather. As early as 1948, Jack Bjerknes, Vilhelm's son, at the time a meteorology professor at UCLA, had worried about the implications. "The everlasting shortcoming of the synoptic analysis by rocket pictures alone lies in the fact that no quantitative picture of the pressure field is obtained," he lamented. This new category of data would be visual. "The analyst must rely on the visible components of meteorology to ascertain to some usable degree the synoptic weather situation."

And what satellites could see would look different. Humans had always looked *up* at clouds; now meteorologists would need to understand them from above. "The change-over to 'looking down' upon the clouds means that the dominant features which serve to identify types of clouds when observed from the ground are no longer to be seen," fretted the RAND authors.

But at the Weather Bureau, a young research meteorologist named Harry Wexler was enthralled by satellites' potential, qualitative or quantitative. Wexler had already demonstrated a knack for being in the right place at the right time—like a meteorological Zelig, as his biographer, James Rodger Fleming, has put it. He was a hotshot from the start, with a mathematics degree from Harvard and a doctorate from MIT, completed under Carl-Gustaf Rossby, a former Bjerknes assistant (and one of the young men in the photograph of the Vervarslinga på Vestlandet in Bergen). After joining the Weather Bureau, Wexler spent the summer of 1940 at the brand-new LaGuardia Field in New York City, bringing the meteorologists there up to date on Bergen School methods—the better to provide forecasts for the high-profile Clipper flights that had begun operating across the Atlantic. During the war, Wexler led research and development for the weather division of the US Army Air Forces—the American counterpart to Weather Station Kurt's engineers. He was the second person to ever fly into a hurricane, riding in the jumpseat of a Douglas A-20. And he was present at the Trinity atomic test, tasked with installing recording barometers to measure the pressure wave of its explosion. He had a keen eye for meteorology's current limitations—and future possibilities.

Wexler had a theory about how weather observation needed to evolve. Up until the technological developments of World War II, meteorology had been limited to two "eyepieces," as he called them: the "microscopic," or what could be seen with the naked eye, which was perhaps twenty miles in good weather; and the "macroscopic," enabled by the observation network. Since the arrival of the telegraph, the macroscopic had been the dominant tool. "By making observations of the weather simultaneously at a large number of places distributed over as large a part of the earth's surface as possible, transmitting them immediately to a central station, and there preparing from them a map showing the existing weather over a particular region, it is possible to estimate how these existing conditions will move and change and hence what weather will be experienced in different areas in the immediate future," Wexler wrote in 1947, summarizing the whole project of weather forecasting as it had stood for a century. Observe, collect, map.

But the limited quality of this view was obvious to Wexler. There had never been enough observation stations—and, really, there could never be enough. The macroscopic view relied "on the 'representativeness' of the observations at one station," he wrote. It was always just a place on earth, a dot on the map. But if the macroscopic view was painfully limited in resolution, the microscopic, in contrast, was limited in expansiveness. The microscopic shows the "smaller and more intricate details which comprise weather," like a photograph of a storm cloud, but "reveals to the meteorologist only a few threads of the large fabric of the atmosphere." Only the macroscopic could chart the "grand features of atmospheric processes." What was missing, Wexler concluded, was a combined view—they needed a bigger picture at a higher resolution.

In 1954, an Aerobee rocket launched from White Sands crash-landed with a surprise: the clear image of a tropical storm swirling in the Gulf of Mexico. *Life* magazine treated the snapshot like a celebrity baby, giving it a full-page spread. Wexler, by then director of meteorological research at the Weather Bureau, enthused about the potential of this new technology: "No one had suspected from sparse meteorological evidence available along the Mexican border that above the earth's surface was a small, intense vortex, perhaps approaching hurricane intensity."

The picture became the starting point for Wexler's further imaginings. In a lecture at the Hayden Planetarium in New York that year, he unveiled a painting he commissioned. The artist had rendered the view a camera might have from space, filled with notable (if imaginary) weather: a line of storms over the eastern United States, fog off of California, a cyclone over Alaska. The German astronomer Johannes Kepler had first used the Latin word *satellite* in 1611 to describe the moons of Jupiter. Not until 1936 was it applied to a human-made object orbiting earth. Now spaceflight had fully captured the public's imagination— three years before it became terrifyingly real with the launch of Sputnik—and Wexler eagerly played up both its gee-whiz aspects and the incredible break it represented from previous weather-observation technologies. "Since the satellite will be the first vehicle contrived by man that will be entirely out of the influence of weather, it may at first glance appear rather startling that this same vehicle will introduce a revolutionary chapter in meteorological science," he said. Among those who were impressed was the writer Arthur C. Clarke, who encouraged Wexler to publish the talk in the *Journal of the British Interplanetary Society*. Science fiction was on its way to being science fact.

The new view this promised was breathtaking. Wexler imagined in detail what the planet would look like floating through space, anticipating the "blue marble" photographs that would come fifteen years later. He could picture the "thousands-of-miles-long trade-wind belts" and the "tiny unseen whirls and vortices that couple the motion of the atmosphere to that of the earth." Wexler saw that satellites offered far more than a new view for weather forecasters—they offered a new view for us all. "There are many things that meteorologists do not know about the atmosphere, but one thing they are sure of is this—that the atmosphere is indivisible," he wrote. "This global aspect of meteorology lends itself admirably to an observation platform of truly global capability—the earth satellite." But he wouldn't live to see this vision of meteorological globalism he envisioned. Wexler died of a heart attack in 1962, at the age of fifty-one.

**By the end of the 1950s, the newly formed NASA was launching spacecraft at an** astonishing rate, and this imagined view of earth was becoming a reality. Vanguard II was twenty inches in diameter and twenty-one pounds. It launched in February 1959, carrying an imaging experiment that used infrared photocells to detect the albedo, or reflectivity, of the earth's surface. Here was a working, earth-observing, space-flying "robot." But it wobbled, making its data useless. Six months later, Explorer VI sent down the first television photo of the earth, but the image was impossible to make out; before the end of the year, Explorer VII was transmitting the first clear images. And in the spring of 1960, a Thor-Able rocket (a descendant of the V-2) launched the first experimental weather satellite—called TIROS 1—from Cape Canaveral. An eighteen-sided drum the size of a breakfast table and the weight of a large man, it was covered in solar panels everywhere but its bottom. A weighted cable extended from its body like a yo-yo, to keep it from spinning wildly. It skidded through space in a fixed alignment, like an amusement park tilt-a-whirl, so that its two cameras pointed at the earth for only part of every orbit. Each the size of a water glass, they scanned a five-hundred-line picture in two seconds and transmitted it to a ground station if one was in range, or stored it on a magnetic tape.

When NASA distributed its first images the afternoon of TIROS 1's launch, everyone could see that a new era had arrived. Before the launch, every new rocket had been just a beginning, an intermediary step. TIROS was different: It was a satellite that finally looked back, adding "a new dimension to our abilities to do things on this earth we inhabit," observed the editorial page of the *New York Times*. For "weather experts, today's successful launching held some of the promise that the discovery of the telescope must have held for astronomers in the seventeenth century." The technology soon proved its usefulness: In 1961, a subsequent TIROS satellite spotted Hurricane Carla, prompting the evacuation of 350,000 people along the Gulf of Mexico.

Yet this new technology of the earth-orbiting, earth-watching satellite was also immediately caught between an optimistic ideology of globalism and the terrifying threats of the nuclear age. It was obvious that the technology that enabled it had alternate uses—like espionage, or carrying a nuclear warhead over the poles. The day of the TIROS 1 launch, President Eisenhower made a deceptively simple statement: "The earth doesn't look so big when you see that curvature." But did he mean it in a spirit of togetherness—it's a small world after all—or conquest? What seemed to surprise everyone was the extent to which this new view of the whole earth seemed to *belong* to the whole earth. Yet it arrived coupled with its opposite impulse, the Cold War cleaving of the planet, and the possibility of its annihilation. There was no way to extricate these new weather satellites from the broader geopolitics of the Cold War and the staggering exertions of the superpowers. In a very practical sense, there were few clear distinctions between weather satellites and reconnaissance satellites, or cargo-carrying rockets and intercontinental ballistic missiles. It worked both ways. The military uses justified the meteorological efforts. The military efforts benefited the meteorological uses.

While Wexler and his colleagues at the Weather Bureau were working with NASA on TIROS 1, the CIA was busy with the top-secret Corona program. A camera mounted to a satellite dropped film capsules that were caught in midair by a passing airplane. When Francis Gary Powers was shot down over the Soviet Union in a U-2 spy plane in 1960, the government insisted he was conducting "weather research"—a claim supported by trotting out another U-2, with a NASA insignia hastily painted on the fuselage. Even the original RAND report on weather satellites was itself a companion to a more general look at the possibilities for using satellites for looking back at the earth—i.e., for reconnaissance. Weather tied the world together, but the technology of the weather had the potential to pull it apart. Even knowing that the civilian and military needs have always been enjoined, it's a harsh realization: We learned to see the whole earth thanks to the technology built to destroy the whole earth.

These political stakes energized meteorology, and not merely as a CIA trick. When John F. Kennedy became president early in 1961, he saw the weather as a potential realm of cooperation with the Soviet Union, both for practical and symbolic reasons. In a story recounted by Michael O'Brien in his biography of Kennedy, one rainy afternoon the president was quizzing his science advisor, Jerome Wiesner, about the technical details of nuclear tests, trying to understand their environmental consequences and looking for a way to stop the cycle of test-for-test in which the Americans and Soviets were engaged. How, Kennedy asked, did fallout in the atmosphere return to the earth?

"It comes down in rain," Wiesner said.

"You mean there might be radioactive contamination in that rain out there right now?" Kennedy asked, staring out the window of the Oval Office. This frank and profound realization—we all live beneath the same sky—soon appeared in Kennedy's rhetoric and policy. Global meteorology held a natural appeal for him. It was a realm in which "the superpowers could collaborate, and everybody would benefit while they eyed each other off politically at other levels," as John Zillman, a former head of the Australian Weather Bureau, put it to me. It satisfied Kennedy's scientific ambitions, while complementing the civilian space and military missile efforts. And it was well poised between the new globalism of the burgeoning jet age, and the technological and geopolitical ambitions of the superpowers.

That April, the Soviet Union succeeded in sending into orbit cosmonaut Yuri Gagarin, who became the first man in space. Six weeks later, Kennedy responded. "I believe that this nation should commit itself to achieving the goal, before this decade is out, of landing a man on the moon and returning him safely to the earth," he pronounced to a joint session of Congress, in some of the most famous words of his presidency. But putting a man on the moon was only the first point in a speech on "Urgent National Needs." Point number two was the development of a nuclear-powered rocket, for exploration "perhaps beyond the moon." Point three was $50 million for communications satellites. And point four—now forgotten—was $75 million to "help us give at the earliest possible time a satellite system for worldwide weather observation." The phrase "worldwide" was crucial. It was partly a nod to America's imperial ambitions of global hegemony, but it also showed that meteorologists' dreams of a "perfect system of methodical and simultaneous observations," as John Ruskin had put it, would soon be government policy.

Wiesner commissioned the Norwegian meteorologist Sverre Petterssen—the Bjerknes assistant—to write a report on the potential of the "atmospheric sciences" for the coming decade. Among Petterssen's recommendations was the creation of a National Center for Atmospheric Research, which would be built in Boulder, Colorado. But Petterssen also made it clear that in addition to cooperation at the national level among major universities, there would have to be cooperation at the international level among weather services. In a speech to the United Nations General Assembly in September 1961, Kennedy again used the aspiration of global weather observation to redirect Cold War tensions away from a destructive missile race and toward more productive scientific endeavors. "Today, every inhabitant of this planet must contemplate the day when this planet may no longer be habitable," he said. "Every man, woman and child lives under a nuclear sword of Damocles, hanging by the slenderest of threads, capable of being cut at any moment by accident or miscalculation or by madness. The weapons of war must be abolished before they abolish us." The list of steps Kennedy proposed to counter this threat of annihilation included the signing of a nuclear test-ban treaty and the establishment of a UN peacekeeping force, both lasting ideas. And then, once again, weather got the last bullet point. In a sentence that could have been removed without anyone noticing, he added: "We shall propose further cooperative efforts between all nations in weather prediction and eventually in weather control."

This footnote in political history became a transformative moment in meteorology. In the years after World War II, the International Meteorological Organization had been reconstituted as the World Meteorological Organization, before becoming a specialized agency of the United Nations, along with its sister agencies the World Health Organization and the International Telecommunications Union. As with the UN more broadly, the

WMO would benefit from the American moment. In 1962, Harry Wexler worked with his Soviet counterpart, Viktor Bugaev, on a report proposing a "World Weather Watch." It would be not only "a coordinated plan for the making of observations" but a calculated effort to communicate those observations automatically and systematically, to process them into "analyses and prognoses," and then to distribute them back to "services which desire them." There would be three systems in this system: A Global Observing System, a Global Data Processing System, and a Global Telecommunications System.

The WMO held its congress every four years in Geneva, and at the next one, eighteen months later in April 1963, the idea took hold. "The concept of the World Weather Watch (WWW) was generally commended as an exciting development," the general summary of the congress recorded, in what counts as diplomatic enthusiasm. Meteorologists from all over the world got busy creating tall stacks of paper, hammering out the details. Over the next decade, dozens of World Weather Watch planning reports were published, including twenty-five in 1967 alone. Their titles point to the breadth of the effort to build a global observing system brick by brick. It was similar in spirit to the effort a century earlier in Vienna, at the first meeting of the International Meteorological Organization, but made newly complicated by all the technological tools now at their disposal. Beginning with Report 1—"Upper air observations in the Tropics"—they went on to address every aspect of observation, transmission and processing of data. There's Report 7, "Meteorological observations from mobile and fixed ships"; and Report 16, "Planning of the Global Telecommunication System."

What's astonishing in reading over them is how deliberately the system was designed. More than merely tying together existing national systems, there was a conscious effort by scientists—on both sides of the Iron Curtain, in all corners of the earth—to design an integrated and coordinated apparatus. It would be, as Paul Edwards described it, "a genuinely global infrastructure that produces genuinely global information." Its core idea was open and equal access to weather information, for operational and experimental use. In theory, at least, the only cost of entry for any nation, however small, was the purchase of a teletype machine. And using technology known as APT—for "automatic picture transmission"—anyone with a small and relatively inexpensive receiver could also have access to the latest satellite imagery as well. By 1975, one hundred weather bureaus around the world had the capability. Having an eye in space revolutionized meteorologists' ability to warn of storms, just as Harry Wexler had dreamed.

The only caveat written into the charter was that the World Weather Watch be used for peaceful purposes only. The UN proper might have been overwhelmed by the festering tensions of a world divided between East and West, but the weather diplomats were insistent on the borderlessness of the atmosphere. This was bold of them, given the technology on which they relied. Weather satellites were so expensive that they could be justified only on national security grounds. Mostly this limitation was technological: The innovation they required overlapped significantly with both intercontinental missiles and spy satellites. But it was also political: The jingoistic appeal of satellites was also a function of how they overflew the whole earth, without regard for the borders below—overturning the historical understanding of sovereignty and territory.

Satellites were a global technology offering a global view, but they were owned and operated by individual nations pursuing national goals. Getting the most out of them required cooperation between nations. To be properly calibrated, a satellite needed corresponding surface observations over a wider geographic scope than ever before, and with more consistency in their distribution. "An accelerated effort to acquire additional conventional data is necessary in order to achieve maximum benefit from data from meteorological satellites," the WMO authors dryly noted. In a satisfying feedback loop, the arrival of weather satellites ironically drove the expansion and coordination of surface observation networks. The technological push that had begun during the war to collect weather observations from new realms, using new technologies, had turned back on itself, so that the new tools needed the old stations. It was just the push meteorologists needed to upgrade, rationalize and organize the existing surface networks. By the 1970s, the newly integrated Global Observing System was closer than ever before to the "perfect system of methodical and simultaneous observations" of which meteorologists had long dreamed.

# 5

# Going Around

There are two categories of weather satellites flying today: geostationary orbiters and polar orbiters. The geostationary, or GEOs, orbit in the same direction as the earth's rotation, making them appear motionless in the sky. They provide constantly updated information about a single area of the atmosphere. The polar, or low earth orbiters, known as LEOs, fly low and fast. They circle the planet from north to south and south to north, overflying a different geography with each orbit and cutting a pattern around the globe like an orange peeled with a knife. The LEOs measure the atmosphere more precisely but cover the whole earth less often. Today's LEOs contribute the most quantitative data to the weather models. When it comes to meaningful impacts on forecasting, especially more than a couple of days in the future, they are the champs. But numbers are hard to see, and it's the GEOs, and the dramatic images they produce, that tend to suck up all the air.

The early weather satellites were all polar orbiters, like TIROS 1. Coordinating their coverage meant timing their orbits, so that each spacecraft flew over a different section of the planet at a different time. But the launch of the first geostationary weather satellite in 1966 changed the formula for global coverage. Rather than looking at the whole earth every day or so, a geostationary satellite could see a single hemisphere constantly, while capturing useful data for an area up to eighty degrees of longitude wide. A WMO group aptly known as the Coordination of Geostationary Meteorological Satellites worked to organize the early GEOs into a constellation of overlapping eyes. By the 1970s, the fledgling European Space Research Organization had a satellite looking down from above 0 degrees longitude; Japan's space agency flew above 140 East longitude; and the United States flew two satellites, one over the Western Atlantic and another over the Eastern Pacific.

Today's constellation of weather satellites is, unsurprisingly, much larger. The World Meteorological Organization maintains a database, easily accessible online, which includes nearly two dozen GEOs and nearly a hundred LEOs. But as is often the case with complex and expansive infrastructures, looking at the complete list tends to obscure its most important members. For example, the sky over the Indian Ocean has nine active geostationary satellites, but that includes Kalpana-1, India's first dedicated meteorological satellite, launched in 2002, and only dimly working; and Feng-Yun 2H, the most recent launch of the Chinese Meteorological Agency, which at the time of this writing is still being prepared for regular duty.

For its part, the United States still focuses its efforts on maintaining one geostationary satellite in position above each coast, designated "GOES-East" and "GOES-West," an acronym for Geostationary Operational Environmental Satellite. Once or twice a decade, a newer version of each will take its place, like a smartphone on an upgrade plan. The old GOES is then either repurposed for some more esoteric task (like communicating with Antarctica) or "boosted" to a "graveyard orbit." Occasionally, a GOES fails before its time. But with luck and planning, a backup satellite will be waiting in "on-orbit storage," ready to be maneuvered into position with a few thrusts of its rockets.

This decade, the GOES program has been in the midst of an $11 billion makeover, run by Lockheed Martin. The new version—the third in the forty-year history of GOES—is known collectively as the GOES-R program. During the time when each satellite is being developed, built, launched and readied for operation, it is identified by its concluding letter: GOES-R, GOES-S, et cetera. When it assumes its operational position, east or west, it gets a number: GOES-17, GOES-18, et cetera. This is important, because each new GOES has a fan club. Meteorologists and weather buffs tend to treat it something like a winning quarterback: They check in with it for every weather event, dissect its triumphs and failings, and share its most dramatic images. But like any star, their value is debatable. The current $11 billion project pays for the life of four satellites, the first two of which launched in 2016 and 2018, but the number is still shocking, especially when placed alongside the entire annual budget of the National Weather Service, which hovers around a billion dollars annually. Put more starkly, this one category of satellite costs more to fly than the entire American weather forecasting system it supports. That expense can be seen as a testament to the importance of satellites to today's weather forecasts, but it is also a clue to the American system's bureaucratic complexity.

Anyone seriously trying to understand how America's weather satellites work would do well to start with an org chart and a list of acronyms (which sometimes have their own acronyms). The primary weather satellites are operated by the National Environmental Satellite, Data, and Information Service, or NESDIS, which is part of NOAA, or the National Oceanic and Atmospheric Administration, which is itself a part of the Department of Commerce. NESDIS is a parallel organization to the National Weather Service, or NWS, and it isn't always called NESDIS; sometimes it is the NOAA Satellite and Information Service. If that's all confusing, it gets worse. The division of NESDIS responsible for the day-to-day operation is the Office of Satellite and Product Operations, or OSPO, which was formed by merging the Office of Satellite Data Processing and Distribution (OSDPD) and the Office of Satellite Operations (OSO). That twisted org chart would appear to have consequences: New American weather satellites have been prone to delays, mishaps, and congressional funding cuts, leading to frequent handwringing over a "satellite gap," in which old satellites fail before new ones are ready to take their place. Most egregiously, in 2003, an unfinished NOAA satellite tipped over onto the floor of its workshop, requiring $135 million in repairs. In 2018, the latest GOES (GOES-17) had trouble keeping one of its primary instruments cool in the sun, rendering it useless at certain times of the day and year. The launch of the next GOES was postponed, in an effort to avoid repeating its predecessor's failure. Complex systems have complex problems, but it doesn't have to be this way. The United States isn't alone in the business of weather satellites. Satellites are global observatories. Not all look down on the entire planet, but they all serve global weather models, aiding global weather forecasts.

EUMETSAT, the European meteorological satellite agency, is both a companion and alternative to the American system. While the US system is mired in bureaucratic complexity in both development and operation, EUMETSAT keeps its structure simple. It is an independent organization funded and overseen by the meteorological services of thirty nations. Its four hundred and fifty employees are housed on a single campus and run under a single leadership. They both plan and operate weather satellites, under one roof. For a journalist eager to see up close how weather satellites work, and looking for a straightforward account of their operations, EUMETSAT is a dream. Even their headquarters in Darmstadt, Germany ("city of science"!), is easy to spot: It is shaped like one of their early weather satellites, with a central cylinder and protruding wings. In the gardens outside, large-scale models of their space-borne fleet are lined up among the shrubbery like cocktail tables at a wedding. The satellites are ungainly looking things. Unlike airplanes, whose graceful lines and smooth skins help them slip through the atmosphere, they have delicate protuberances and pocked exteriors. One looks like an engine dipped in gold, another a washing machine with its casing blown off. But it is good to see them there nonetheless. EUMETSAT's real fleet is well out of sight, long ago launched into space. We usually see satellites in one of two ways: as artists' illustrations of what the satellite looks like zipping through space; or half under construction, blown out under fluorescent lights and pored over by technicians in bunny suits. I visited EUMETSAT to look at their satellites in a new way, in the moment when they come closest to earth.

**"You know, people are always criticizing the weather predictions,"** said Yves Buhler, EUMETSAT's director of technical and scientific support, when I met him in his sunny corner office in Darmstadt. A French rocket scientist, he was dressed like it: crisp white shirt, spread collar, breast pocket full of fine-tipped pens. "But globally, it has become much more accurate. And it has become much more accurate, also, in the medium range—so a week, two weeks. Why is that? Because the satellite observations are providing a uniform coverage of the earth. There's no black hole of an area." The global view is everything.

EUMETSAT's first generation of geostationary satellites, known as Meteosat, launched in 1977 and flew until 2002—not the same satellite vehicle but the same generation, like the model version of a car. EUMETSAT's more recent geostationary satellites—Meteosat-10 and -11, soon to be replaced by -12 and -13—covered what Buhler described as "our area." By that he meant Western Europe and Africa. In 2006, they moved their older geostationary satellite, Meteosat-7, to the east, over the Indian Ocean. (It has since been replaced by Meteosat-8.) Just as old cellphones find second lives in the global south, so too do aging weather satellites. "The WMO is very happy that we have a satellite over there to feed a bit better the models," Buhler noted.

The coverage of the LEOs—the polar orbiters—is organized by the Initial Joint Polar System, a collaboration between EUMETSAT and NOAA to coordinate orbits and data formats. EUMETSAT's Metop polar satellites take the midmorning route, overflying each section of earth over that place's morning. NOAA's JPSS polar satellites fly by in each place's afternoon. This duet is relatively new. EUMETSAT's first polar orbiter only launched in 2006, which is startlingly recent. Before then they relied entirely on the American polar orbiters. In the era of weather models, it has only been the most recent generation of satellites that have provided a step change in the quality of forecasts. The slow evolution of the system was part of what makes this whole endeavor of weather satellites feel, in the way of all space technology, always slightly outdated but extremely high-tech. It is all still a work in progress, and it surely always will be. EUMETSAT has a launch and operations schedule carefully blocked out into the 2030s.

**But weather satellites also have daily routines.** Midway through our conversation in his office, Buhler came up short and studied the big watch on his wrist. He spun around to the phone on his desk. "Do you know when the pass is? Yes. That's perfect. That's perfect." Buhler led us through the satellite-shaped building, clicking through electronic locks and shuffling down natural-lit stairwells, saying passing hellos to scientists and engineers in French, English, German and Italian.

Behind a final set of heavy double doors the LEO control room was a wide, tall space, set up like a Hollywood mission control with task chairs, dozens of screens and large ticking countdown clocks mounted high up on the wall. Technicians kept a close eye on EUMETSAT's LEOs and the GEOs from adjacent control rooms. Each room had its own personality and rhythm, like the satellites being watched. The technicians in the GEO control room kept a steady vigil; if everything goes well, not much happens. The LEOs are livelier, and their life is more syncopated. Every thirty minutes one of their LEOs makes a "pass": the period in each orbit when the satellite flies over the North Pole, allowing it to be in radio communication with its ground station. When Buhler and I walked in, Nico Feldmann, a young ponytailed operations engineer, jumped to his feet. "Twenty-three minutes; Metop-B; over Svalbard!" he barked. It took me a moment to realize that he was joking, playing Spock to Buhler's Kirk, pretending we were on the bridge of the Starship *Enterprise*. But then I realized that he was only half-joking. We were really there to meet a spaceship.

Controlling EUMETSAT's polar orbiters requires a fiber-optic connection across Europe and under the Barents Sea to Svalbard, the Norwegian island above the Arctic Circle. From there, a radio connection is made from a dish antenna ten meters in diameter, sheltered by an egg-shaped dome the size of a cottage. The dish that serves EUMETSAT is one of thirty-one set on a plateau known as Platåberget, next door to the Svalbard Global Seed Vault, which stores seeds from around the world in case of an apocalypse. While Buhler, Feldmann and I chatted in the LEO control room in Darmstadt, the antenna in Svalbard rotated on its spindles, quick and smooth like a robot arm, until the massive bowl was aimed at precisely the point on the horizon where Metop-B would appear, rising like a speck of dust in a sunbeam.

Feldmann and his colleagues call that moment "AOS," a spaceflight acronym meaning "acquisition of signal." Metop-B orbits the earth fourteen times a day, flying nearly north to south, and then south to north (at an angle, or inclination, of 98 degrees), aiming its instruments downward through a narrow swath of the atmosphere each time. A polar-orbiting satellite, by definition, overflies the poles every orbit; but, with the earth spinning beneath it, it crosses the Equator at a different longitude each time. Each orbit around the earth takes 102 minutes, but the satellite is only visible to the ground station on Svalbard for anywhere from twelve to fifteen minutes of that, depending on the time of day, because Svalbard is not located precisely at the North Pole. On that day, the shortest pass would come during the Norwegian night, around two or three o'clock in the morning, when the satellite would be headed for daylight on the other side of the globe. The main task of each pass is to download the gigabytes of observational data typically collected while the satellite is flying around the planet. Technically speaking, this is called a "full dump," and it is kind of like trying to download a movie over your neighbor's Wi-Fi while driving by their house. (Except it works.) "You need to get the data down, and you need to get it down fast," Buhler said. To reduce the delay between the satellite's observations and the distribution of the data, there is also a "half dump," when the satellite flies over McMurdo Research Station, in Antarctica.

Each pass brings a mix of drama and routine. It was why I was interested in this control room and not the one next door. The geostationary satellites are just that: stationary. They appear to float above us, sloth-like in their stability, keeping a watchful eye. That is an illusion, of course; they are actually flying through space at around three kilometers per second, completing an orbit of the earth once each day—in other words, at the same pace as the planet itself. But GEOs are always in touch. With the LEOs, each pass has some excitement. If anything went amiss while the satellite was out of range—if an instrument malfunctioned or any temperature or voltage parameter went out of limits—an alarm would sound. Feldmann gave his own analysis of the geostationary spacecraft managed by his colleagues next door. "GEO is boring," he said. Buhler was more diplomatic. "It's always interesting to watch what's happened over those hundred minutes when the satellite has been flying, invisible to our site," he mused.

Metop-B's approach to Svalbard from the far side of the earth was indicated by a red-numbered LED countdown clock on the wall. "The first activity associated with a pass is usually twelve minutes before, when we establish the connection to the ground station," Feldmann said. The moment approached. We waited. A machine chirped. "There it is," Feldmann said. "Now we have twelve minutes to send commands to the spacecraft."

"And to get the data down," Buhler added, pointing his finger up. We all watched as a column of boxes on the monitor turned green. "And the telemetry looks . . . nominal," Buhler said, relieved, using the space lingo for "normal." "Telemetry" refers to the basic health of the satellite and its systems, things like temperatures and voltages. For each pass of the satellite, the two key values Feldmann monitored were the duration of "TM" and "TC," standing for *telemetry*, meaning the health data received from the satellite; and *telecommand*, indicating the ability to send commands back. Those transfers happen over relatively low S-band frequencies. The juicy stuff—which is to say the "science data"—comes over X-band, which is a higher frequency microwave band. Feldmann pointed at another column of green rectangles. "If those are green, it means science data is coming down." We watched as the numbers ticked up, bit by bit. Feldmann went down the list of acronyms of the different instruments: ASCOT. GOME. GASSS. IASI. AMSU-1. AMSU-2. Buhler quietly mouthed their names along with him, like a dad at a spelling bee. By this point, we were 1.8 gigs into the dump. If the data being downloaded were a movie, it was one only some experimental filmmaker of the future could imagine: It consisted of 10,000 channels of infrared and radar soundings, shot from space through the clouds. There were five minutes left in the pass.

As Buhler, Feldmann and I talked through the details of Metop-B's routine, I could appreciate the satellite as a busy working instrument, watching the atmosphere as attentively as any earthbound weather station. It tucks away its images in its solid-state memory banks and then zaps them down to the surface. The data itself isn't a snapshot, a single click in time, but more like an unspooling filmstrip, as the high-flying robot traverses the atmosphere, pushing through space with its nose pointed down like a bloodhound.

Before I'd had a chance to notice, Metop-B had finished its dump. "Now you see the science data has been downlinked," Feldmann said in his best David Attenborough voice. "We have just a little bit over one minute left to send it commands." But we had nothing to tell the satellite. Everything was green, everything was *nominal*. In another part of the building, EUMETSAT's computers had already begun to send the observations out to the world through their data links, paying special attention to the most eager customers: the operators of the weather models, hungry for the latest measurements of the atmosphere. There was a pleasing symmetry to it: The satellite sucked up its observations of the whole earth, and EUMETSAT sent them back out to the whole earth. As Eisenhower had recognized, this view of the planet belonged to the planet. Polar orbiters were the game-changing observatories of today's forecasts; they were too small to see with the naked eye, but I now had a new vision of how they circled above.

For its part, Metop-B was back below the Norwegian horizon, a happy robot, flying alone, doing its thing. Beside the clock counting down to the next pass was the spacecraft's odometer: the number of orbits completed. On that afternoon, Metop-B was in the midst of its 10,754th revolution around the earth, since its launch in 2012. In less than an hour it would come back over the horizon. Its rhythm is tied to our lives, to the spin of our planet that defines our hours.

I bid goodbye to Feldmann.

"We're here all day," he said.

# 6

# Blasting Off

When Vilhelm Bjerknes first attempted to calculate the weather, he was so eager for more observations that he would often collect them himself. As his son Jack recalled, in addition to the "tapping of the typewriter in my father's study," "heard year in and year out," during summer vacations they would launch "impressive kites, far greater than toy ones," laden down with recording instruments. I wondered what he would have thought, a hundred-odd years later, in the exhibit hall of the annual meeting of the American Meteorological Society. Its outer edges were a low-key bazaar, filled with small booths belonging to mom-and-pop weather-instrument makers, specialty publishers and university meteorology departments. But the center was a bonanza of high-tech observations, dominated by large booths belonging to the titans of the military-industrial complex, companies like Northrop Grumman, Ball Aerospace, Harris and Raytheon. They had plush carpeting, uniformed staff and halogen spotlights aimed at scale models of their wares: drones and satellites suspended by transparent threads or ensconced in niches, like statues in a museum.

Along with the flagship geostationary and polar orbiters—the LEOs and the GEOs, the Metops and the GOES—there is a third category of weather satellite. These spacecraft have narrower, often experimental, missions, to push the technological limits of space-based observation. Since they compete with each other for funding, they often have snappy names—like CALIPSO or CloudSat—the better to pull hundreds of millions of dollars out of government budgets. They look not only back at the earth but toward the future, testing and refining new kinds of instruments, like lasers in space that scan the tops of clouds. In any given year, there is an alphabet soup of these new earth-observing satellites preparing for orbit.

Among the displays of the American Meteorological Society exhibit hall, held that year in Atlanta, one in particular intrigued me. Known as SMAP, a name it shared with a Japanese boy band, it had an unusually narrow function: to measure soil moisture from space. Soil moisture is a funny data point in meteorology. The weather models include it as a variable—but an infrequently updated one, in part because it is so poorly measured. SMAP promised to change that, with two orbiting instruments doing the work of ten million fixed ground-based sensors. It felt to me like a project Bjerknes would appreciate: a bold and audacious attempt at intensified observation of the earth, in the service of a better calculation of the weather. And it was an example of the tail of the weather models wagging the dog of the entire observation system. It wasn't a new observation that the weather models might use, it was an observation the models already wanted.

"You go to space because you want a global map," Dara Entekhabi, the project's lead scientist, said to me, when I tracked
him down in the cavernous hallways of the Atlanta Convention
Center. Born in Iran and educated in the United States, Entekhabi was a professor at MIT in the department of civil and
environmental engineering. He had started his academic career
in geography before moving into engineering, keeping, as he
said, "one foot in meteorology and one foot in hydrology." At
the time, in the 1990s, they were two separate disciplines, in two
separate departments, one in earth sciences and one in engineering, and the two rarely interacted. As far as meteorologists were
concerned, rain ceased to exist once it hit the ground. Hydrolo-
gists, in contrast, didn't care where the water came from. As a
young faculty member at MIT, Entekhabi looked at the connections between rain and water on the ground, especially as it was
represented in soil moisture. He knew the weather models could
benefit from the data, especially since it acted like a new variable.
Models often improve when their resolution is increased, and the
atmosphere is observed and calculated at finer and finer scales.
But increasing resolution requires increasing the computational
load. There are more observations to be calculated—or, in the
modelers' language, "resolved." But that increased precision can
bring greater turbulence to the weather model, like the flicker of
a pixel when you zoom in on a screen. "After a time you're chas-
ing your tail," Entekhabi said. An increase in the resolution of
one parameter demands the increase in resolution of the other,
and vice versa. Since soil moisture is omnipresent in the models
but remarkably absent in the observations, Entekhabi saw an opportunity.

He spent fifteen years refining his justifications for the method of collecting global soil moisture data with an $800 million satellite. His case evolved from a one-page statement of the "science imperative" and the "technological readiness"—i.e., that the world needed this satellite, and it could be built—to a four-hundred-page document, modestly called "the handbook." Key to the project's green light were the constituents beyond weather modelers, eager for SMAP to exist—most notably, the Department of Defense. Soil moisture affects both low-level fog forecasts and the calculation of density altitude, which is a crucial metric for calculating the performance of aircraft, particularly in the mountains. (Most famously, the crash-landing of a helicopter during the raid on Osama Bin Laden's compound in Abbottabad, Pakistan, may have been caused in part by a miscalculation of density altitude.) This desire for better soil moisture measurement was crucial in getting SMAP off the ground. As has always been the case with satellites, the military dollars always came in far bigger stacks.

Each summer, Entekhabi and his team would go out on field campaigns, mounting their prototype instruments to the fuselage of a small plane, then calibrating the readings against simultaneous ground measurements of the soil moisture. Their satellite had a poetic name to start—Hydros, god of the waters—but soon became known as SMAP, an acronym that stands for "Soil Moisture Active Passive." To cut through the bureaucratic process, the satellite had to tell its own story. The "active" and "passive" refer to SMAP's technological special sauce, its combination of a radar and radiometer. A radar sends out radio waves and receives their echo; a radiometer only receives them. The combination allows an unusual combination of breadth—with global coverage repeated every two to three days—and accuracy, with the radar allowing for better measurements than the radiometer alone might. But the name was above all tactical. "We wanted it to be extremely clear just in the name, so when the folks at NASA headquarters are sitting in a closed room making decisions, there's no question that it's active and passive, and that it measures soil moisture," Entekhabi said. The project hung on by a thread—"I don't know how many times we got canceled"—but its case was compelling.

SMAP was assembled beneath the bright lights of the most famous workshop for American spacecraft: High Bay 1 at the Jet Propulsion Laboratory in Pasadena, California. This gymnasium-sized room is the place where the machines that had gone farther from earth than anything else, ever, had been made: missions to Venus, Mars, Jupiter, Saturn, Uranus and Neptune. It was easy to be blasé about this achievement, especially for my generation, which grew up with the launching of the space shuttle a matter of routine (until we saw it explode—a true puncturing of innocence). But the longer I spent thinking about how we look at the earth from space, the more my sense of amazement increased. The guts and carcass of the nearly finished satellite lay in the middle of the white-tiled room, protected behind retractable stanchions, surrounded by equipment racks, tool carts and computer workstations on wheels. It was really going up *there* for the eccentric purpose of measuring soil moisture back down *here*.

That gee-whiz unbelievability was the theme at the Jet Propulsion Laboratory. The main guard booth had an enormous curved window like an astronaut's visor, with a sign that read "Welcome to Our Universe." In the centralized visitors' waiting room, a TV played not CNN but NASA-TV—where, as I waited, I watched a video of a Russian Soyuz spacecraft docking at the International Space Station, an image that would have astonished and pleased President Kennedy.

JPL was founded in the 1930s, when a Hungarian-born professor at the California Institute of Technology named Theodore von Kármán first tested an early rocket engine in a dry canyon nearby. Army money gushed into the lab during World War II, when engineers studied the stolen designs of the German V-2 rocket, desperate to replicate its capabilities. By 1947, they had developed a primitive guided missile, known as Corporal, a necessary first step toward sending rockets into space with any hope of deliberate control—which was an essential step before you could put a nuclear warhead on the tip. With the missile race on and the space race about to start, JPL was at the center of the enormous effort to solve all of the technological problems that arose: not only rocket basics like aerodynamics and propellant chemistry but also radio communications, new types of instrumentation, supersonic wind tunnels and the unprecedented process of building anything of such high quality that it could be confidently flown "beyond reach of repair," in the rocket engineers' wonderful phrase of art.

Their progress was astonishing. Within a generation, JPL engineers had leaped from building primitive rockets to spacecraft flying to other planets—like the Mariner series that in the

1960s and 1970s, visited Mars, Venus, Mercury and Saturn. Yet JPL was haunted by the space program's fundamental duality: They would create the machines that would open up a new era in human exploration—truly out to the edges of the solar system—while also creating the technologies that could destroy humanity. There was no excising the genes of war from the DNA of weather satellites. We put things in space for many reasons, but we learned to put things in space for only one.

SMAP, with its own dual military and civilian mission, was no exception. A press minder buzzed me through a security gate and we walked across the campus, dense with desert-tan buildings and well-tended shrubbery. High Bay 1's small viewing gallery was up a narrow flight of stairs, like a skybox in a stadium. Sam Thurman, SMAP's deputy project manager, met me there. With his crew cut and white shirt he looked like he stepped out of a NASA photograph from the 1960s, and talked like it too.

SMAP—splayed beneath us, behind a plate-glass window—was in the final stages of testing. It was medium-size by spacecraft standards, "about the size of a Mini Cooper," Thurman said, in the bemused, dad-jokey jargon typical of NASA. SMAP had a lot of moving parts that often moved in unusual ways. It had a distinctive profile: A circular antenna reflector extended above the main part of the spacecraft like the halo of an angel costume. The reflector was made of a light mesh, and it was big: twenty feet in diameter when it had opened up like an umbrella after launch. It would spin at 14.6 revolutions per minute—or one revolution every four seconds—expanding the coverage area on the ground in the same way you might twirl a flashlight to illuminate a forest path.

Almost everything in it was custom designed, with Northrop Grumman, Boeing, NASA's Goddard lab and JPL all contributing bits. It was "a double-spinner," Thurman said, "meaning that one piece of it spins and one piece of it doesn't." Some parts, like the slip rings that transmit signals between the parts that spin and the parts that don't, were something Boeing had learned how to build for communications satellites. But these were a one-off. "You don't go buy a radar system for measuring back scatter at a certain wavelength of the L band," Thurman said. Every component of the satellite was "very expensive to design, very expensive to build, very expensive to test." Thurman explained the reason for that, although it should have been self-evident: "Once you shoot this puppy off, you can't get it back."

SMAP, like all satellites, was delicate and robust, like a racing bike. No expense could be spared, but compromise was constant. That season, a year before launch, everything was being tested: "shaken, cooked, frozen and fried," Thurman said. SMAP would be wheeled across the JPL campus to a shaker table, which would shake the satellite to simulate the forces of launch. When testing was complete (and assuming none of its $30 million components crumbled off like plaster), SMAP would be wrapped in a special shipping container and trucked up the coast, escorted by California Highway Patrol, to Vandenberg Air Force Base, where it would be mounted to the pointy end of a rocket.

In the days leading up to the launch I found myself thinking about the global view satellites provided, and the political baggage they carried. To make arrangements to see SMAP blast off, I had been communicating with the public affairs office of the 30th Space Wing at Vandenberg. The base had a long Cold War history. While its unofficial slogan was "Nuke Free since '63," it still hosted tests of the Minuteman III intercontinental ballistic missiles frequently enough that the neighboring communities were nonplussed by a fireball streaking across the sky. Yet the base still treated each missile launch as an event, which seemed curious. If there were an actual nuclear war, the country's four hundred and fifty active Minuteman III missiles, scattered across Wyoming, Montana and North Dakota, were supposed to launch within fifteen minutes—a notion as terrifying as it was improbable. Seeing the efforts at Vandenberg, it didn't add up. Either the tests, with their weeks of preparation, were being treated with an abundance of caution; the Minuteman fleet was a far more elaborate and expensive operation than could be grasped; or the fleet was only symbolically useful, a modern Maginot Line that would barely fly at all. Most likely there was some truth in all three.

The thing about space I kept coming back to was how much imagination it requires. Satellites are not merely "beyond reach of repair," they work far beyond our sight. The view from a satellite has changed the way we think of the earth, inspiring what the philosopher Peter Sloterdijk calls "a Copernican revolution." Just as Copernicus showed how the earth was not the center of the universe, spaceflight and the images that have returned have created an "inverted astronomy," in which we now often imagine our planet as if looking down on it from above. But it is harder to look back at the camera. We "struggle to engage with satellites because they lie so firmly beyond the visceral worlds of everyday experience and visibility," as the geographer Stephen Graham has pointed out.

That difficulty spills over into the weather forecast. To "check the weather" is a totally banal activity; we may as well be turning on the stove or flushing a toilet. Yet when we check the weather our mind's eye travels widely in space and time, rising up above the earth to look down at the clouds with radar or satellites, while zooming ahead into the future with the forecast. Seeing the SMAP launch, I hoped, would make the system more tangible, less imaginary.

When I arrived at Vandenberg I parked at the main gate, just outside the town of Lompoc. If it weren't for the military police with their sidearms and helmets, the sign announcing "Force Condition Alpha" or the white truck with "Cryogenic Maintenance" painted on the side, I might have mistaken the place for the entrance to some vast winery. Along with a dozen or so men and one woman, all weighed down with tripods and camera bags, I boarded an old-fashioned white school bus driven by an enlisted man in fatigues, who guided our press group slowly across the darkened base toward the launchpad.

I saw it first through the eucalyptus trees: a structure as tall as a skyscraper, brilliantly lit up by spotlights, with a blinking red light at the top. SMAP would go to space on a Delta II rocket, 370 of which had been launched successfully since May 1960. It was an impressive number, especially given the complexity of the effort and the excitement that still surrounds it. The preparations to "mate" the SMAP "observatory," as it was known, to the Delta II had begun six months earlier. It was scheduled to launch the next morning. That evening was the "rollback," when the service tower at "Slick 2"—as the launch complex was known—was rolled away, leaving the rocket standing on its own. But the technical purpose was only part of the point. A rollback is treated with ceremony and superstition, like the rehearsal dinner of a wedding.

Our school bus pulled up alongside a low ridge, just beside the launchpad, and we scampered up for a better view back toward the rocket. While the others were busy setting up camera equipment by the glow of their headlamps, I stared at the spaceship we'd all come to see. Giant spotlights illuminated it from three sides, catching dust in their beams. Around me were fuel canisters, yellow safety banisters, white SUVs and a flatbed truck mounted with a row of video cameras. But there were also plovers in the surrounding dunes and a three-quarter moon (unvisited for more than forty years). Diesel generators thrummed, and the air was filled with a disorienting scent of ocean and kerosene. I had been following SMAP's progress for more than a year; I had seen the EUMETSAT's polar orbiter pass by and tracked the upgrades of the new GOES satellites. But still I was surprised by the enormity of the effort of launching this instrument into space. This was obviously about more than measuring the moisture of the soil; it was an exercise in prestige, a less categorizable attempt at scientific advantage, and the manifestation of a compulsion to keep doing this—because to halt the project begun sixty years ago would be, for NASA, an unthinkable admission of defeat.

A white Buick sedan, shiny in the night, pulled up to the foot of the ridge, and everyone's attention shifted toward it. An astronaut emerged: NASA's administrator, Charles Bolden, who'd traveled into orbit four times on the space shuttle. By coincidence, it was NASA's Day of Remembrance, timed to the anniversary of the *Challenger* disaster. Administrator Bolden had started the day at Arlington National Cemetery, laying a wreath for NASA's dead, before flying across the country to Vandenberg for the launch. Our small crowd of scientists and technicians who had worked on SMAP, and their families, pushed in around him on the dusty hillside. He kissed a baby and took a question from a second-grader: "What will the satellite do?" the child asked. Bolden described SMAP's purpose, ticking off the bullet points of its raison d'être, "but the big thing we're trying to do is understand the planet," he said.

It all seemed quite close to the spirit of JFK's announcement, which had set NASA on its greatest run. It raised a question, and I pushed forward to the front of the group to ask. How did SMAP fit into that broader project, of the human use of space and its exploration? "If you look forward twenty, thirty years, we're trying to get humans on Mars," Bolden said. "Everything we do today is really one more little step toward getting there." Then he struck another note, more wistful than I had expected.

"I came to NASA in 1980, and honestly, when I came here I thought we would be much farther along than we are now," he said. "I am one who believes that when we lost *Challenger* we lost decades. People say we lost our sense of risk taking. I don't agree with that. I just think it took us time to get the nation to heal, and to get people willing to take risks again. Not the agency, but the people who support us: Congress, the administration. Even now, when we talk about going to Mars, a lot of people just think that's a bridge too far." SMAP was, in that context, easy. It was one more component of the vast machine, clicked into place. The view from space was an incredible technological leap, made not so long ago. Would another leap come again? And what new technological possibilities would *it* open up? How deep into the atmosphere could we see? And how far into the future would that reveal?

Bolden got back in his Buick, the crowd dispersed and the project managers' children all went home to bed. The giant scaffold began to pull away from the rocket, so slowly at first I had to stare to see it moving. The tower itself was gangly, dripping with cables and ugly orange lights, while the rocket seemed to suck up all the light. Having been assembled inside the tower, no one had seen it exposed like this, ready to leave the planet.

**Very early the next morning, I rode the Air Force school bus again to a clearing in** a eucalyptus grove, a safe distance away from the launchpad. At Vandenberg's mission control, the launch team had been at their computers ("on console," they say) since past midnight. Beginning then, liquid oxygen was loaded into the rocket, final navigation instructions uploaded and the countdown begun.

Rockets from Vandenberg rise into what's known as the "western range," an expanse of ocean and sky stretching across the Pacific. In the final moments of the countdown, it had to be free of obstacles—clear for "sea, land, air and space," as the Air Force colonel responsible put it—whether small yachts or foreign destroyers. The Air Force's global tracking system was activated to track the rocket, another wildly expansive piece of Cold War infrastructure. Ninety-nine seconds after launch, the Delta would jettison its solid rocket boosters, which would then fall into the ocean. Forty-one minutes later, after peaking at a speed of 12,000 mph, the rocket would arrive at a "parking orbit." Fifty-seven minutes after launch, the SMAP observatory would separate from the rocket. But for all that to work so that the proper sun-synchronous orbit was achieved, the launch had to happen within a three-minute window. SMAP was going to blast off beginning at 6:20 a.m. and forty-two seconds, or it wasn't going to blast off at all.

As it happened, SMAP didn't launch that day, or the next. The first morning, strong winds high up in the atmosphere canceled the launch—a "scrub." The space press assembled in the eucalyptus grove groaned when the announcement came over a loudspeaker and immediately started to pack up. The next day, a mechanical issue stopped the countdown before we got back out to the grove. I had to get home. I watched SMAP leave earth perched on the foot of my own bed in New York, via a NASA video feed. The pad was enveloped in a heavy fog, and it took me a minute to register what I'd missed: three seconds of rumble before the Delta disappeared into the clouds.

The bad news came five months later. On July 7, 2015, SMAP's radar stopped working. NASA engineers spent a few months testing and fiddling, until exhausting their options and declaring the instrument "lost." SMAP's radiometer still worked—the "passive" instrument to the radar's "active." But the resolution of SMAP's soil moisture maps would forever be reduced. "There's a nonzero probability of failure," Entekhabi had said to me. "It's a tough environment. All those things have to work. It is a very risky thing. I've waited for it for a long time, so I'm hoping . . ." His voice had trailed off.

Another American satellite had failed. The cliché is that "space is hard." But I understood it another way: The earth is hard. Understanding the atmosphere required coming to terms with many moving pieces, conceptually and mechanically. I felt a sense of awe at the apparatus of space, but it was tempered by the reality that the adjacent project of observing the atmosphere would never be complete. The complexity was too great, the resolution never fine enough, the preponderance toward chaos always a nick away.

For its part, SMAP's mission was saved from an unlikely corner: the C-band radar on the Sentinel satellites, launched by the European Space Agency for climate change research. Its similarity to SMAP's broken radar was enough that scientists could combine the data sets, nearly re-creating SMAP's intended sensitivity and resolution. Rather than two instruments working in tandem on one spacecraft, it would be two instruments on two spacecraft, their readings algorithmically combined into a single map of the earth's soil moisture. It was quite the hack, and a perfect example of the ways in which the weather machine was a system of systems: instruments riding on satellites, mathematically woven together back on earth, to create a cohesive digital model of the atmosphere. Achieving meteorologists' dream required not a single vision but an amalgamation, a stitching together of thousands of instruments. Only combined could we have the picture of "the present state of the atmosphere" that Bjerknes envisioned.

The next step was to calculate its future state. For that we needed the weather models.

# PART III

# SIMULATION

# The Mountaintop

The heat in Colorado was a dry mountain heat, broken in the afternoons by wild thunderstorms that rose up behind the Rocky Mountains. On a cloudless morning in June I drove up out of the city of Boulder, to the point where the suburban houses gave way to sandstone rock formations rising out of meadows of pine and prairie grass. There, at the end of a sweeping driveway, was a cluster of copper-hued towers with crenellated tops, like a mountain fortress. The buildings' scale was difficult to discern. They had long slit windows and a sturdiness that made them seem like a natural outgrowth of the mountain.

This was the Mesa Lab, the spiritual home of American weather science. Its opening in 1966 as the flagship of the National Center for Atmospheric Research (NCAR) had marked a moment of reinvention for meteorology, sparked by President Kennedy's interest. The amazing technological advancements of the Cold War brought entirely new realms of potential to the field. New satellites could look down on the clouds. New electronic computers could calculate equations. New radars could see storms over the horizon. "The sky is quite literally the limit," Walter Orr Roberts, the first director of the National Center for Atmospheric Research, said when the building opened. "No field of science—even atomic energy or medicine or space exploration—offers a greater potential for the good of all mankind than does the field of atmospheric science"—as meteorology was renamed by its boosters (and funders) in Washington.

Roberts wanted the architecture of the new lab to reflect that ambition. He hired I. M. Pei for his first major commission and asked for a building that expressed "both the contemplative and exciting aspects of scientific activity." It should be "monastic, ascetic, but hospitable." It should have "soul." For inspiration, Pei camped out on the site among the deer and rabbits, and he visited the Anasazi cliff dwellings in southwestern Colorado. He specified concrete mixed with pink sand from a nearby quarry and had the walls hammered in a way that made them rough-hewn, like stone. The burly rose-colored towers frame the sky, linking the permanence of rock and the capriciousness of clouds. I could feel the tension Roberts envisioned: between stasis and change, what is and what will be. The Mesa Lab is a timeless building devoted to new ideas, which is why I was there: to consider the gap between the sky and its human understanding; between what we know precisely about the atmosphere and what we can't; between the weather of the present and the weather of the future.

In practical terms, I wanted to know how the weather models worked—how they took the observations and turned them into a forecast. I had seen weather infrastructure all over the world, in space, in the sky and pushed out to every corner of the map. But the models were the suns at the center of this solar system, holding all of that other infrastructure in their orbit. Their appetite for data dictates how and where new weather observations are collected. Their ability to provide automated forecasts for any point on the globe powers the weather apps most of us use. The models are the achievement of Bjerknes's vision, the source of the daily forecast, the engine of the weather machine.

And they are hard, meaning difficult. As a rocket scientist at the Jet Propulsion Laboratory put it to me when I explained my broader project, landing a spacecraft on Mars requires dealing with *hundreds* of mathematical variables; making a global atmospheric model requires *hundreds of thousands*. "That's complicated!" he said. But I wasn't willing to dismiss the whole enterprise as beyond comprehension. The stakes are too high for that. This technology isn't a new gadget—some better smartphone or talking speaker—but a globe-spanning system of increasing import. It helps that the weather models are not assembled in secret by a corporation but out in the open, collaboratively, among scientists and government agencies from all over the world, albeit so slowly that their construction has mostly gone unnoticed.

A weather model has an anatomy, a clear and logical separation of parts. A model needs *observations* of the weather; it needs to know what the weather is, to know what it could be. A model needs what is most easily called *physics*—a set of equations to describe how the atmosphere evolves (as first described by Bjerknes). And a model needs to put the two together with *computation*, which Lewis Fry Richardson tried and failed to complete at the Western Front but which is now most often handled by a supercomputer. The success of any model depends on the strength of each, like a three-legged stool. How good are the weather observations coming in? How well is the model able to mathematically calculate their behavior over time? And how quickly can the computer make those calculations?

There are two main categories of weather models: "experimental" models, developed by scientists to focus on certain problems, like how clouds and rain (or hurricanes) form; and "operational" models—meaning those run by weather services for the purpose of everyday forecasting. At the Mesa Lab and the National Center for Atmospheric Research more broadly, they don't predict the weather, at least not as a matter of routine. Their knack instead is for the big picture, for how the parts fit together.

At NCAR, one of the scientists with the biggest picture is Jeffrey Anderson. We met in the lobby, amid the fading posters with pictures of lightning and clouds. His training was in meteorology and computer science, but his work had spanned software engineering, applied mathematics and statistics—all with an eye toward improving weather prediction. Now in his fifties, he had a narrow face and extraordinary eyebrows like Abraham Lincoln. Along with seemingly everyone else in Boulder, he was tanned and fit, as if ready at any moment to bound up the mountain outside. Along with seemingly every other man in the building, he wore khaki pants and a blue button-down shirt. Walter Orr Roberts used to brag that the Mesa Lab was a maze—"There are twenty different ways to go from my office down to the chemistry laboratory," he said—and I raced to keep up with Anderson as we went up the ladders and down the chutes of stairs and hallways to his office in one of the towers. A recumbent bike was set against a wall. I felt like a student visiting a professor during office hours—especially when Anderson wasted no time disavowing me of my premise.

I had come with an assumption, which turned out to be a misunderstanding: that the weather models are another of those things that we lazily refer to as "algorithms," computer programs where you put one kind of data in and another kind of data comes out. I expected that weather observations from all over the world flow into a supercomputer, and the supercomputer works like a meat grinder to transform that data about the present into forecasts of the future. But that wasn't quite right. Or, rather, it didn't capture what was happening inside the supercomputer, and why it all worked so well.

"In your description there's this idea of the tension between the simulation model and the real world," Anderson began, tenting his hands in front of him and speaking slowly, like Dr. Falken in *War Games.* "But what sits between those, which you didn't mention, is data assimilation." The weather observations of the present do not *become* the weather predictions of the future. Instead, the atmosphere inside the model is a going concern. It exists continuously, a planet inside the machine. The weather of the real world is "assimilated" into the models, matching up the outside atmosphere with the simulated atmosphere. It is as if the observations are *correcting* the model's earlier forecast, like a ballroom dancer still learning the steps. The whole process is, as Anderson put it, "nontrivial," but it's the secret of every weather model's success.

Each click forward into the future can be thought of as a

given weather model's hypothesis, ready to be tested against reality. "The scientific method is: We're going to predict some observation," Anderson pointed out. Except weather models aren't constrained to the present and the near future. They can test their hypothesis, their particular way of calculating the weather, using the entire archive of weather data. "You can develop a new prediction tool and test it on five years of data. But then you still have fifty-five other years of independent data to check it on, in addition to the fact that if you just sit on your hands for a year, you're going to have a whole new year of stuff that's completely independent," Anderson said, with an obvious thrill. "It's really hard to fool yourself!"

By definition, the best weather models are the ones that most closely simulate the atmosphere at a given moment and over time. One challenge is that whereas the conditions in a model are organized and presented perfectly rationally, snapped to a grid of the modelers' own design, the conditions in the real world—as observed by instruments—are going to fall where they may. The weather station in Utsira is where it is (and long has been); its location will never be changed to accommodate the data structure inside the computer. Making things more interesting, there is plenty we don't know about the present weather. There are places without observations; and places with observations that are wrong. The observing system may be vast, but it is imperfect. The glory of good data assimilation is that it allows for the model to compensate for places where observations are sparse. It becomes a bridge between the areas that are well observed and the areas that aren't. The surprising result of that discrepancy between model space and real space is that the model, you might say, is more detailed than reality—or reality, at least, as it is observed.

Anderson's explanations changed my mental image of the model. Rather than a meat grinder that transmogrifies the weather of the present to the weather of the future (a one-way process), I now pictured two spinning earths side by side. I saw the real earth, the planet we live on, in the view we have gained from traveling to space. And I saw the model earth, its simulated atmosphere swirling with clouds and storms, with the bonus capability of being able to run in fast-forward, into the future. The secret of a good weather model is how well it makes the two match up.

Nullschool, a popular website built by a former Microsoft engineer named Cameron Beccario, makes this point well. It's a beautiful visualization of the earth's winds swirling around our familiar blue marble. But it's not entirely correct to see it as a visualization of the earth's winds as they are observed; rather, it is a visualization of the earth's winds as they are modeled. The fact that we conflate observations and models is a testament to how good the models have become.

Yet the models' trick of matching the simulated atmosphere to the real atmosphere is imperfectible. That sounds bad, but it means models will always be getting better. There is no cap on our ability to measure the atmosphere; there can always be more observations. There will be no future moment when the atmosphere's behavior is perfectly understood. And computing, as we know well, continues to improve. When we think about improving the weather models, it is tempting to plead toward one component of the system or another. If only we had a perfect image of the atmosphere, thanks to better observations! If only we better understood how the weather "worked"! If only we had the fastest computer ever conceived! But the best modelers know that the three legs of the stool—observations, physics and computation—are always working together. The supercomputer is at a loss without the physics that describe the calculations. The calculations are useless without the reams of observation data. The observations are overwhelming without the computers to parse them. Improving weather models (much less understanding them) requires coming to terms with all three.

Weather models are complicated, but they're certainly useful. "There's lots of fields that have simulation that don't actually make predictions, and there's lots of fields that have observations that don't actually make predictions," Anderson pointed out. But in weather prediction "you have to put together the observations and the simulations—because people really want to know what it is going to be like tomorrow, over and over and over again." That distinguishes weather modeling from other kinds of predictions, like those for elections or sporting events. Weather simulations are uniquely ubiquitous. Not only can scientists constantly tweak them, but the appetite for their improvements is broad and unending. We consult the forecast every day, and we check its accuracy viscerally, zipping up our jackets or wiping raindrops from our glasses. Tomorrow becomes today, and we know right away if the forecast is right or wrong. And the better the forecast gets, the more we want it.

From the perspective of the mountaintop in Boulder, the benefits of all that demand are obvious. "Numerical weather prediction has been a sixty-year story of continual improvement,"

Anderson said. "It just keeps getting better, and there's no particular reason to expect that that's going to stop."

And nowhere did weather models improve more, and more consistently, than at the crest of another hill, in Reading, England.

# 8

# The Euro

The bus stop, not so far outside of London, was called "Weather Centre." I pushed past the schoolchildren with their bright backpacks and stepped off in front of a compound surrounded by a sturdy steel fence, like an embassy. It was cold and misty. I walked up to the empty gatehouse, pressed the intercom button and looked into the camera. The gate clicked. I started up the long, winding driveway toward a ring of twenty-two flags on tall poles, belonging to the twenty-two member nations of the European Centre for Medium-Range Weather Forecasts (ECMWF).

When Lewis Fry Richardson imagined what it would take to calculate the weather, he pictured 64,000 people filling a grand stadium, their brains humming away in parallel to solve the equations he had worked out, which would take the weather of the present and move it forward into the future. He believed his forecast factory should be surrounded by playing fields, mountains and lakes, so that "those who compute the weather should breathe of it freely."

ECMWF, in Reading, England, is, in every practical sense, the realization of his dream: a forecast factory that supplies the world with accurate predictions "faster than the weather advances." In place of Richardson's 64,000 human calculators is a pair of Cray supercomputers, whose hulking cabinets fill two rooms the size of volleyball courts, like the stacks in a library. They are among the fastest supercomputers in the world, and they're upgraded every two years to stay that way. At the time I visited, they had 260,000 processor cores, capable of conducting 90 trillion calculations per second; together they weighed more than 100 metric tons, ingested 40 million weather observations a day and ran calculations at a sustained rate of 90 teraflops. Their dark cabinets were screen-printed with illustrations of European monuments, like a Starbucks wall.

But all those numbers will surely have changed by the time you read this. What won't change—because it's the keystone of the European Centre's success—is the way the "compute time" of the supercomputers is used. ECMWF assigns 50 percent of its computer resources to research, so that the scientists in the building can easily run complex experiments. They can think up a new way of calculating some element of the atmosphere's behavior, try it out in the model and know—as surely as the next day's sky—if it actually works better. The benefits of this are immediately apparent to the scientists working in the building, many of whom come for a few years at a time, on loan from the weather services of the European nations that fund the place. They can test changes to their piece of the code, and they have the horsepower to effectively rerun the model using their slightly tweaked version of the numerical atmosphere.

Just as there are two main categories of weather models—experimental and operational—there are two main scales of models: regional and global. Those focused on a single region can simulate things like cloud structures more finely, often resulting in better rain and snow forecasts. With less data to crunch, regional models update more frequently, as often as every hour. They can also work in smaller units of time, predicting the changes in the weather for fifteen-minute increments, or recently even less. As with an old-fashioned paper map, there are advantages to looking at a smaller slice of the earth, in more detail; but, as with a paper map, too much detail can be unwieldy. High-resolution models require more computing power, which is expensive. And, for forecasts further into the future, short-term precision has diminishing returns when it comes to accuracy. Small errors compound, sending a forecast off in chaotic directions. Global models, in contrast, allow for a greater expansiveness, in space as well as in time. Global models are the heavyweights of the weather world, pushing the boundaries of accuracy for longer-range forecasts.

As for which weather model is the best, there is little reason to beat around the bush. The current and undisputed champion is the flagship global model operated here, at ECMWF, officially known as the Integrated Forecasting System, or IFS, but informally called "the Euro" (especially by Americans). In the closely watched horse race of models' "skill," the Euro is king. Compared to the world's other global models—like the ones run by the Met Office in the UK or the National Weather Service in the United States—the Euro is the most accurate the furthest out in time (if sometimes only slightly). It is also the most improved, the most often. A common statistical benchmark for modelers (but nonsensical to those of us who forget our umbrellas) is known as the "500 hectopascal anomaly correlation," and according to this esoteric measure the graph of ECMWF's performance climbs like an airliner on its way to cruising altitude—except it hasn't gotten there yet. It is still climbing, for two decades now and counting.

**The idea for the European Centre emerged in the late 1960s, along with the Euro-**pean Union. Crowded in by the astonishing exertions of the two superpowers, moving deliberately and diplomatically with the hope of assuring the long-term prosperity of Europe, a handful of countries began poking around the edges of what they might do in the realm of scientific and technical research. Their early list of possibilities was noticeably peaceful, polite and international, like a list of grievances posted in the corridor of an Olympic dormitory. Among the pan-European problems they considered tackling were "language translation," "annoyance caused by noise" and "refuse disposal." But in the category of "Scientific and Technological Research," two problems were pushed forward: "Longer-term weather forecasts" and "influencing the weather." The "influencing" part, although very popular at the time, was dismissed as impractical; we still can't change the weather (at least, not in a controlled way). But longer-term forecasts seemed a good problem to tackle.

From the perspective of a burgeoning European community, meteorology had a lot going for it. Its scientists were used to working internationally, having long been aware that the atmosphere formed its own union. Numerical weather prediction demanded serious computing power, of a cost and scale that overwhelmed budgets, particularly for small nations. And it was politically safe, in a way: There was no possibility of abject failure, only middling success, and any improvement in the weather forecast could be considered a win. A plan came together to focus on the medium range, leaving short-term forecasts to the individual weather services of each country and long-range forecasts to the future. ECMWF was structured to be independent, with funding from its member nations, its own governing council and a laser-focus on its founding mission, of the "medium range," meaning, at the time, three to five days. (Their own success has since expanded that definition.) The supercomputer would be the biggest available. The best and the brightest from each nation's weather service would cycle through the place. And the results would be far greater than an individual weather service could hope to achieve. That was the vision—and it's worked.

The European Centre for Medium-Range Weather Forecasts officially opened its current building in Reading, England, in 1979, with a 999-year ground lease from the British government—written with the eyebrow-raising proviso that the building be handed back, at the end of that time, "in its original condition." On opening day, Prince Charles ceremoniously received the forecast for the Ascot races, which proved to be mostly correct.

At the time, the longest range forecast worth the dot-matrix printer it came out on was about three days. By 2005, ECMWF's model was reliably "skillful"—meaning its predictions were more likely to be right than the average temperature for the date—five days out. One could make a forecast from the historical averages, but the goal is to be always better than this. By 2015, ECMWF's scientists had squeezed out another day from the future, which meant the six-day forecast was now as good as the two-day forecast in 1975. Then they moved the goalposts: By 2025, ECMWF wants to have a model capable of predicting high-impact events two weeks ahead. (It predicted Sandy eight days ahead.) This is the truly remarkable thing about the place: not merely that ECMWF had the best global weather model in the world but that it had been constantly improved, for forty straight years.

"It's amazing, how this excellence works," Florence Rabier told me. At the time she was director of forecasts, but she was soon promoted to director-general, the first woman to hold the job in ECMWF's history. "People come here because they want to be at the best place in the world, and they don't want to lose that. If you start showing a slight decrease of performance or somebody is catching up with us on some parameter somewhere, they get all annoyed. Really, it should be me being annoyed, because I'm responsible for the performance of the forecast. But it's really down to the scientist doing the individual piece of code that really takes the responsibility. They are incredibly motivated by it." Rabier has always been one of those motivated scientists. She came to the European Centre in 1996 from Météo-France, the French weather service, with a project to implement her pioneering data-assimilation technique, known as 4dVar. Originally part of her dissertation research, it is now a keystone of the European model's superiority.

We were talking in ECMWF's busy cafeteria, over a proper

lunch of roast chicken, hot rolls, rice and salad. The elaborate spread has become a cliché in the technology world, but by all accounts this cafeteria has always been the hub of the place. The long narrow room was striped with rectangular tables in unbroken rows, and everyone seemed to fill one seat beside the next, like passengers on a lifeboat. The scientists were young. Norwegians, French, Serbians, Italians, Irish and everything in between, all dressed similarly but looking different: men and women, tall Scandinavians and bespectacled Moroccans, blond-haired or brown-skinned, all wearing the academic's uniform of jeans and brown shoes, the women in sweaters and the men in sweaters over dress shirts. Many were on a multiyear assignment from their respective national weather bureaus. Rabier stood out, dressed like the boss in a navy suit jacket, offset by shoulder-length auburn hair. These were the authors of the algorithm. For all but the people who cycled through this room, the model was a black box. But they were the ones who had painstakingly assembled and improved it, drawing upon the best research from around the world. What that meant, though, was that they weren't predicting the weather on a daily basis but, rather, working month by month and year by year to improve the program that did the predicting.

The cafeteria had been busy in the morning. The cafeteria had been busy at lunch. The cafeteria was busy again in the afternoon. There were two high-quality automatic coffee machines, one with watery fair trade, the other with dark roast espresso. I fished a porcelain demitasse from a dishwasher rack and moved to the back of the scrum of scientists pushing toward the coffee robots. The scientists filled their cups and headed back to their rectangular tables, squeezed together like army cadets, leaving neither an empty seat nor a sliver of silence, as the low winter sun shone between the clouds and through the big windows.

If the European Centre's original impulse was to merely share a "computer installation" and exchange weather data, the early international spirit had clearly seeded a longer-lasting ambition: "We want to be the best," Rabier said. When I commented that they already were, she dismissed the idea. "Well, yeah, day-seven is as good as day-five twenty years ago," Rabier said of the model's skill. "But it's still not as good as day-one. We're always pushing the limits, wanting more. That's why I think we always feel, 'Well, it does improve, but it's not perfect.'" She shrugged. "It will never be perfect." Her job, as she saw it, was to maintain the bureaucratic infrastructure of their improvement. Coffee service aside, the scientists' collaboration was carefully conceived, with scientists from the forecasting and research divisions constantly comparing the model's capability with its possibility.

What makes the model better? "It's always a mix, you know," said Rabier's colleague Peter Bauer, the head of ECMWF's model division. He was near fifty, tall, lean, German. He wore a giant silver diving watch and a fitted black shirt and black jeans. "Sometimes people tend to simplify and they think, 'Oh, if we only had the perfect observation network,' or, 'If we only had the perfect model.' But you have to work on all fronts similarly." The space agencies, like EUMETSAT, for example, are always eager for justifications for more observations—investing millions, if not billions, to get them, with new satellites. "But sometimes— not always, but sometimes—the actual bottleneck is not having additional observations," Bauer said. As often as not, the bigger bottleneck to the model's improvement is properly assimilating the observations already being collected and matching them up properly with the world inside the model. The more the scientists can improve the data assimilation, the more usable information can be extracted from the observations. The better the data assimilation, the smaller the corrections the model needs to make. But that process can be slow. On average, each year ECMWF was adding observations from five new instruments—not satellites—into the model. In 2013, data from a total of fifty instruments was being assimilated. By 2018, that number had grown to ninety.

The complexity was again staggering. There seemed to be this endless list of challenges in building the models: better observations, more observations, better use of observations, more efficient use, better calibration, higher resolution, higher accuracy, faster computers, or more frequent outputs. There was never one thing to tweak. Every time I thought I might have a handle on how things worked, I would hear about another layer.

For example, a model that can make the best use of available observations is a great first step toward creating good forecasts. But one of the modeler's key areas of inquiry is the behavior of the atmosphere *between* observable data points. There are things going on in the atmosphere that are, as Bauer put it, "fundamentally unresolved" by the model. These are physical processes whose behavior is "parameterized," in modeling language, meaning they are calculated based on averages within a grid space rather than a value at a specific point. Contemporary models have a range of distinct "parameterizations" that define the behavior of the atmosphere at a scale smaller than the base grid of the model, which is itself getting tighter every couple of years as computing power increases.

At ECMWF, these parameterizations, known as "schemes," have a human component. For the most part, each of the different schemes is managed by a single person at ECMWF, whose job it is to improve its ability to predict how the weather (or one aspect of it) will evolve. One scientist works on radiation, another on clouds; there is convection and turbulence—both in the "free atmosphere," meaning up in the sky; and in the "boundary layer," meaning close to the surface of the earth. But when I say "worked" what I really mean is "improved." The scientists at ECMWF painstakingly and ceaselessly reassemble and refine the model itself, testing new methods to see what best matches the sky. The process is both iterative and inherently experimental.

"People become efficient very fast, and their level of satisfaction is high," Bauer said. "They say, 'Hey, I run experiments with the ECMWF model, and I'm testing my little science change, after two weeks!'" For an atmospheric scientist, it's a concrete motivation. "A paper is nice, of course," Bauer joked, "but if you do something in an operational context like this, that's quite good."

That might seem like an obvious strategy, but a few months before, at the American version of ECMWF—NOAA's Center for Weather and Climate Prediction—I saw how challenging it could be to create a structure that allowed new ideas to be tested on the model. A group of scientists visiting NOAA for a modeling workshop had a simple request of the model's masters: They wanted to "check out the code," like a book from a library, so they could have the chance to fiddle with it, in what they hoped would be a productive way. But even doing that—the first steps toward experimenting with improvements to the model—precipitated a long discussion among them of technical hurdles, like the system's security requirement that anyone who logs in needs to have a fixed IP address (which is not something that your typical home Internet connection has). Without an established method, trying out modifications on the working model was like trying to figure out how to change the tires on an eighteen-wheeler without pulling over to the side of the road. The scientists were left to work on improvements using experimental models, which made it harder for them to incorporate them into the operational model. (They were also prohibited from using government funds for catering, even at meetings—so, no lunch.)

**The European Centre is housed in a compound of modernist beige-brick buildings,** diplomatically bland, arranged around a courtyard nearly in earshot of the M4 motorway. In the center of the courtyard was a dry fountain, heaped with rubber duckies. There were duckies of all types: Some wore the native dress of a European nation, others the logo of a multinational corporation. I spotted a soccer jersey and a Sigmund Freud. I first understood them to be a geeky joke that had perhaps gone on too long. You could now buy an ECMWF ducky from the receptionist. But in time I saw how they were also a fair sign of the spirit of the place: open, welcoming, international, ready for new ideas, a bit relentless.

Just inside the main entrance, off a hallway that connects the cafeteria with the supercomputer, is the Weather Room. For years it had been set up like the common room of an Oxford college, with clusters of club chairs, books and journals. But a recent renovation had added a large wall of screens, installed edge to edge in a grid six wide and two tall, programmed to cycle through the model's outputs in the form of maps.

Every weekday, an analyst was on duty to watch the screens, looking for extreme events, unusual features in the (simulated) atmosphere or big differences between the European Centre's model and the models of other weather services. "Ooh, we have some strange waves there," Rabier said, as an example. "Is that realistic or is it noise in the model?" The analyst job rotated, but the week I was there it belonged to Tim Hewson, a large guy in his forties, with the bearing of a rugby player. He'd only recently arrived at ECMWF. He'd spent his career at the UK Met Office, rising to the position of chief forecaster. In a weather-obsessed place like Britain, being chief forecaster was a bit like being poet laureate. Many could claim some skill, but his was officially certified. It was a testament to the prestige and challenge of the European Centre's project that he would come here next—not even to forecast in the traditional sense but to use his experience to improve the outputs the model produced. On a daily basis, he had no say as to what went out to the world. Instead, he was like a trainer monitoring the vital signs of a star athlete, thinking up new ways of improving their performance. What *was* the model doing? What was the weather doing? The goal, above all, was to sketch the gaps between them, so that the world inside the model—present and future—might better match the real world outside.

Hewson had no shortage of second opinions. This was by design. He was stationed for the week in a glass-enclosed office, almost like a booth, with a task chair and a couple of computers. In a place filled with so many weather lovers, the scientists often stopped in to look at the big maps on the screens or to have a meeting at one of the little coffee tables spread with satellite brochures. Hewson paced back and forth between his workstation and the big screens along the wall, scratching his chin, making notations on a white notepad, listening to the opinions of whoever came through. Each day, he would put up a report on the European Centre's internal wiki, or message board. Everybody was welcome to comment. On Fridays, there was an open meeting to discuss the week.

But each calendar quarter ECMWF hosted a bigger meeting for reflection and discussion, known as "FD/RD," that brought together the Forecasting (F) and Research (R) divisions, which together comprised nearly the entire staff of the Centre. I'd timed my visit for the last one of the year. It seemed a good opportunity to see how the scientists actually made the model better. What would they ask each other, and how would they answer? Of particular concern was a recent upgrade completed by the UK Met Office, two hours down the tracks in Exeter, which was good enough to put it within striking distance of the European Centre's superiority—a situation that was, you might say, motivating. "I am sure that the first question on everybody's lips is, 'So are they catching up with us?'" Rabier said. "Sorry, but we are not slightly competitive, we are *very* competitive."

It was easy to see the place in *Top Gun* terms: The European Centre was the best of the best. It had the smartest scientists, the biggest supercomputer, the most focus and determination. But that impulse glossed over two key points: The first was that, on a day-to-day basis, the differences among the major global weather models was slight. Certainly there have been occasions—like Sandy—where one model, usually this one, gets it right sooner. But the more crucial point was the strategy at the heart of the Centre's identity, the thesis of the place, one fully infused with the crux of what makes weather models work, broadly speaking: ECMWF had the best weather model in the world because its model was always getting better.

**The next morning, I filed into the lecture hall upstairs from the cafeteria along** with everyone else. The day was split in half: First, the forecast division would present the overall performance of the model, including its "headline score" and how it handled recent interesting weather events. Then in the afternoon, the discussion would turn to what developments were coming next for the model— and what should be coming after that.

The ground was always shifting beneath them. Since all their competitors were also always upgrading, they tracked two kinds of improvement: making a better forecast (which almost everyone is doing, year by year) and making a better forecast than everyone else. When the models' "anomaly" scores—a measure of how right the model is—went up on the big screen in the lecture hall, a scientist behind me let out a "Yeah!" like a stockbroker on the trading floor. No, the UK Met had not caught up. "The lead is still clearly there," Thomas Heiden, an Austrian scientist, assured the assembly. "Considering that we didn't have a cycle change for a year it still looks good." There were nods all around. I wouldn't want to say they wished failure on others, but they did enjoy their successes.

When a Moroccan scientist named Mohammed Dahoui stood at the lectern to present for the data-assimilation group, he went down the list of which observations—from which satellite instrument, or which category of the global observing system— were turned off and turned on, according to how much they helped the model. In that calendar quarter, the most interesting development was the model's first use of data from a Chinese weather satellite, FY-3b, known as Fengyun, or "wind cloud." This was particularly remarkable because the satellite had launched five years earlier—making sure the data was useful to the model had taken that long. The process was tedious, but if its additional observations led to even the slightest improvement in the scores, it was worth it. It wasn't unusual to run an experiment to see exactly how useful. In particular the Fengyun instrument that measured humidity had been highly anticipated—pending, as an ECMWF report put it, "a detailed assessment of the data quality of these microwave radiometers." Making a weather model required much more than buying a new supercomputer. The scores only improved thanks to the fine-toothed comb the scientists here applied to every tangled strand of the system.

Sometimes the opposite happened: An instrument would disappear, and the models' scores wouldn't suffer. That was the surprising result that quarter, following the outage of certain instruments from the US satellite system. After close analysis, the data-assimilation team could not find any evidence that the loss of observations degraded the quality of the forecasts—especially compared to the other major global models. It was a startling point: The US satellites had functionally gone offline, and the European model didn't notice? A scientist in the front row guessed that the Euro's 4dVar system (the one that Rabier had developed) had allowed the model to be more resilient to the dropping out of a particular set of observations.

After the instrument report, Ivan Tsonevsky, a Bulgarian scientist with dark hair in a Caesar cut, clicked through some major weather events of the last quarter from all over the world—not merely in the European nations that funded the place. With a global model, it doesn't matter if a weather event takes place in Buffalo, New York, or Annapurna, Nepal. Except that when it snows three feet in the mountains above Annapurna, as opposed to the suburbs of Buffalo, there may not be an official snowfall observation recorded. In that case, the model's last forecast becomes a proxy for ground truth. It's one of those moments when the model has more data than real life. The snow is really there, but the only record of it is simulated.

In the back-and-forth between the scientists, their questions were always some variation on "How does the software behave?" How can we better understand the complexities of this thing we've built? How can that understanding help us rebuild it better? How does the model handle this? How does it represent the extremes of weather that matter most to us, like cyclones or massive snow dumps coming off the Great Lakes? A bad weather forecast is really just a moment when the model's version of the atmosphere diverges from the reality that came to be. Every scientist in the room kept a piece of the model in their head. Each presentation was followed—and sometimes interrupted—by questions. Alan Thorpe, the director general of the center at the time, had no hesitation jumping into the intellectual fray, nor did the younger scientists have any compunctions about arguing back. "Yes, but" was their most common reply. Their accents and complexions varied, but they all huddled together over this apparatus—an idea, as much as a thing, embodied in software, running again and again on the two supercomputers down the hall.

After lunch, the group reconvened in the European Centre's fabulous 1970s boardroom, which was decorated with wall tapestries showing weather systems rendered in purple tufts of fabric. Fifty scientists sat around the council table, as if it were a graduate seminar. For two hours, they were all completely attentive. No phones were out, no one moved. The debate was how quickly to push for a higher resolution model. Rabier argued that "the users"—the national weather services—demanded it, in part to drive their regional models. But higher resolution didn't necessarily lead to better forecasts, and it came with costs to computing time and complexity. For the moment, the issue was left unresolved. The resolution would eventually be upgraded, of course—but at that moment there were plenty of smaller steps to take toward the model's improvement. With the sun low, all the scientists went back to their offices to do the work—after a stop in the cafeteria.

I had spent the day listening to the scientists discuss the year-by-year improvements of the forecasting system. But there was a second axis of time at the Centre: the model's twice-daily "runs." Meteorolo-

gists and their computer systems around the world have structured their days according to the European model's schedule. If
a big winter storm were bearing down on the northeast United
States, it wasn't unusual for them to stay up late to wait for the
output and let it guide (if not define) their forecast for the next
twelve hours. The European Centre had staff on duty around the
clock to monitor the computers and telecommunications, but it
wasn't as if they needed to press the start button on each model
run. The cycle repeated on an automatic schedule and rarely, if
ever, did anyone stop to watch it. Yet each run took hours, which
seemed long enough to note each step if you did.

Adrian Simmons agreed to try. Having joined in 1979, Simmons was one of the longest-serving scientists at ECMWF, and
he had worked on nearly every piece of the model in his time.
He was the unabashed wise man of the institution—seemingly
everyone's mentor. We hatched a plan to borrow an office (he
was semi-retired), log in to the supercomputer, and Simmons
would narrate the progress of the model run as best he could.

When the clock struck four the first box silently turned
green: "Getting Obvs," it read. Simmons, wearing a tan zip-neck
sweater and a check shirt, sat at a long desk with a standard-
issue black workstation. The bulletin board above his head was
pinned with historic weather maps and diagrams of model process flows—the precise thing we were about to watch. If this
were Richardson's forecast factory, runners would be frantically
grabbing the sets of temperature, pressure and humidity readings as they came in from all over the world via telegraph. The
human calculators would be flashing their illuminated signs at
each other as they worked out their equations. But that's not how

our digital world worked. Outside the window was a stand of trees, but the supercomputer itself was in the next building over, down a flight of stairs and at the end of a long hallway.

For the past twelve hours, the most recent observations had been flowing into it through the fiber-optic cables that snaked up beneath the driveway. The system's first step was to move them from the storage servers to the supercomputer. Having collected all that raw data of the present state of the global atmosphere, it then got to work, comparing the observations for the last twelve hours with the forecast for the last twelve hours—the real world with the simulated world—and adjusting the model to match. Here was the dance I'd been hearing about, the pas de deux between the model and reality, one leading, the other following. The present was the key to the future, but the way the European model worked, it also had to compare this observed version of the present with the model's version of the present—aka its most recent forecast. There was one version of the atmosphere inside the model and another version of the atmosphere observed by the instruments, and it was at this step in the process that the two had to be compared and adjusted.

Fortunately, that wasn't usually so hard. "Our twelve-hour forecast is pretty accurate," Simmons said, "so the observations only tend to make relatively small corrections. It's not like our first guess is the weather we had last summer, or the weather we had on the same day last year. If it was that, then we would really struggle to get this method to work well, because we would be making large changes to the observations. But because we have a good starting point—because we've already assimilated all the information we have from the observations up to twelve hours ago, there's a lot of constraint already in there." The model was an excellent bootstrapper. It was so good a model of the atmosphere, and ECMWF was so capable of assimilating observations, that the adjustments necessary between the afternoon and evening observations were quite small. You might say that the model already knew what the observations were going to be—because it's good at predicting the future. That was the beauty of data assimilation: It predicted the next set of observations. Which made it easy to make small adjustments to keep them on track.

All this happened at relatively low resolution because the process is, as Simmons put it, "expensive"—meaning it takes a lot of processing power. In a system of ECMWF's scale, "computational budget" is everything; its billions of steps need to be carefully arranged so they are completed soon enough to be useful. (As Richardson had warned: "Perhaps some day in the dim future it will be possible to advance the computations faster than the weather advances.")

Having collected the observations, the model then began working through its myriad interlocking components—things like surface temperatures, surface humidity and snow depth. Some things were simpler for the supercomputer than others. Sea surface temperature, for example, was borrowed wholesale from the Met Office rather than analyzed directly. One model's version of the earth system informed another's, forming a constellation of models that shared specific data products and depended on each other for the work. "It sounds a bit competitive today, but it's not uncooperative," Simmons assured me. It was models all the way down.

As the status indicators flipped through their colors, Simmons furrowed his brow. "Now it's gone a bit cryptic," he said. The model was working through something listed as "satid224," which indicated a particular instrument on a particular satellite. A process called "variational bias correction" began to run, which adjusted the satellite observations on the fly. (The office in which we were sitting was where much of the technique had been developed.) The observations themselves were in flux based on the character of the instrument and the atmosphere it was observing. They required constant calibration and adjustment. That startled me. If they were slippery, did that mean there was no ground truth? Was anything absolute?

"There are some observations we don't correct," Simmons said. "There are some observations we *believe* are the truth, and they basically anchor the system." It was a funny way to put it, I thought, that the scientists *believed* certain things and built their simulated earth on that foundation. But it was a good reminder that the model was a model—not reality, nor its mirror, but a representation. When it came to the infinity of the atmosphere, the model would always be a bit of a guess.

That reminded Simmons of another crucial characteristic of this system: The weather observations it contained, all the hundreds of millions of them, were not merely observations of a value but observations keyed to a particular location on (or above) the earth. The model was defined by its three-dimensional grid, to which the observations, of course, didn't necessarily adhere. A weather balloon floating through space measures continuously, tracing its three-dimensional path—rather than only at the abstract points in the sky that correspond precisely to some model somewhere. It used to be that the observations were transmitted only with a WMO station ID number—as at Utsira—and the computer would go to a separate table to look up the location, according to latitude and longitude. That evolved with the satellite era, in which a single instrument might traverse the earth in nearly its entirety each day. The observation stations were no longer fixed in place. Satellites freed weather instruments from the previous bonds of space. To boot, all those observations weren't merely in space, they were in time as well. The fourth dimension was always present.

Once the initial conditions were set, the model started off into the future. While previously it was comparing its past forecast to recently observed conditions, now it was running ahead of reality, crunching through the descendants of Bjerknes's equations. It was just after five o'clock in the afternoon in Reading, but inside the supercomputers it was six o'clock. "The six o'clock is a real forecast," Simmons said. "We are into future time. It's an hour ahead of *now*."

This felt like watching a magic trick being performed—if not explained—and it looked like installing a new piece of software, watching the progress bar inch along. Things had slowed. A reasonable person might say this demonstration, the length of a feature film, was lacking in drama, and I was getting hungry. "Nothing's gone red yet," Simmons murmured, a little disappointed that everything had gone smoothly. As the model progressed into the future, we chatted about airplane delays and Christmas presents. I could hear raindrops on the roof of the building. A cleaner rolled his cart down the carpeted hallway. Simmons checked his watch. "I can last until day ten, and then I've got a couple of things at home."

At 5:42 p.m., Greenwich Mean Time, we were two hours into this adventure, and the model had time-traveled four days into the future. At 5:50 p.m., we hit day five. Things were speeding up. The farther the model went, the less it had to do. It was tuned to be less precise as things went on because the forecast was less reliable anyway. After day six, the model would only write an output for every six hours. We watched as it took forty-five minutes for the model to compute the first five days, while it was doing the last five days in under half an hour. By the time we crossed day ten, Simmons had forgotten about his chores at home. "Watch carefully now as it's going to finish," Simmons said, barely containing his excitement. The final scripts stuttered down the screen, like an all-text Space Invaders. The super-computer emptied its cache, cleaned itself up and got ready for its next run—ten hours later.

"Okay, that's finished now. Oh, there we are. We're done," Simmons said. He checked his watch again. "It's just gone quarter past six." It had taken the forecasting system two and a quarter hours. Richardson's dream had come true.

Simmons reminisced: "In the early days we would go down and sit and watch it with the operators, if it was a really big change. Because if something went wrong we'd have to fix it. These days you can fix it from home." He sighed.

That evening we were the only ones keeping an eye on every step, but there was a whole world out there on tenterhooks, waiting for this output. The supercomputer down the hall was working up a model of the atmosphere, then spraying it out to billions of us, all over the world. There were meteorologists waiting to parse the latest run, and there were computer systems ready to ingest the latest forecasts into their own models and send them out to all of us. Simmons began clicking around, plotting some maps to see the big weather trends in the coming week—the state of the atmosphere across the planet, in the ten days until Christmas. For a weather geek, this was like eating the finest fish, fresh out of the sea. I wondered if there was any value in having this information first.

"The energy markets," Simmons said. Since energy futures trade not on the weather itself but on the forecast, it was theoretically possible to use what we were looking at for arbitrage. "But we all have jobs," he said. "Well, I don't have a job, but I'd still like to be let in the building."

Simmons needed a few minutes to answer emails, and I saw myself out. I walked down the narrow hallway, past the scientists' offices, each with its light wooden desk, its task chair, its desktop computer wired into the big computer another fifty steps down the hall. Each office held the body that held the mind that worked out the processes, made of math and silicon, fed by all the flying satellites, buoys and balloons. This was the seat of our weather capability. This was the accumulation of forty years—or a hundred and fifty years—of cumulative effort. This was the source of the forecast.

So then where did it all go?

# 9

# The App

The first weather forecast appeared on the Internet on February 23, 1991, put up by a gangly graduate student at the University of Michigan named Jeff Masters. The atmospheric science department in Ann Arbor had a satellite dish on the roof that collected a data feed from the National Weather Service. For an assignment in a class on interactive weather computing, Masters wrote a short program that gave the data feed a simple interface: Anyone on campus could type in an airport code, and the National Weather Service forecast for that city came back. At the time, Michigan was the heart of the Internet, thanks to the work of MERIT, a nonprofit that held the government contract to operate the Internet's backbone. With the help of their staff, Masters bolted a second little program onto his original one, plugged a spare machine into their network, and made his weather tool available to anybody on the Internet.

Very quickly, that became everybody on the Internet. In the first week it was live, Masters had five hundred users. Three months later, 120,000 people a week were getting in the habit of asking for the forecast by three-letter code. He named the tool Weather Underground, in a cheeky allusion to Michigan's radical 1960s political group, because at the time it was radical: "a geeky, underground, cutting-edge sort of thing," as he put it to me. Within a year, Weather Underground was one of the most popular sites on the Internet.

But if Masters's site was a new way of getting the weather forecast, it was the same old forecast. At least in those early days, all Weather Underground did was take the outlooks written by the National Weather Service meteorologists and sent out over their existing distribution network, and put them on the new Internet. In 1995, Masters and a few colleagues spun off Weather Underground as a for-profit company, but they narrowly missed their chance to register the weather.com domain.

The Weather Channel—a cable TV darling in the United States since the 1980s—got to it first. But they soon realized they had a problem. Unlike Weather Underground, they were writing their own forecasts for television, and their impulse for their new site was to just let the TV guys with grease pencils fill in the same data. But that wasn't working as planned. The web audience was global, for one thing, and it operated on its own schedule. The Weather Channel needed more places updated in less time. For help, they went (as I did) to the National Center for Atmospheric Research in Boulder.

At the time, Peter Neilley was a scientist there, working in a lab focused on the practical applications of research. Sandy-haired and rosy-cheeked, he had been interested in meteorology since he was a kid in New Jersey in the 1970s, eager to forecast snow so he could go skiing—a pragmatism that later guided his choices in graduate school at MIT. While his classmates chased a more theoretical understanding of the weather, Neilley built his own computer and customized an operating system to help move the department's research data from analog to digital. "I was always interested in how we can make a better forecast for tomorrow," Neilley said.

He wasn't invited to the Weather Channel meeting, but he could hear it going on from his own office across the hall. They were planning a so-called "expert" system, which would make a programmable logic out of the methods of human forecasters from around the world. Neilley knew it wouldn't work. In what he calls his life's "flap of a butterfly moment," he grew exasperated enough with the overheard conversation that he burst into the meeting and told them they were thinking of it all wrong. The expert system wouldn't scale. There were too many forecasters around the world, with too many individual approaches, and their methods were changing along with the data sources on which they relied. How could they keep forecasts up to date for everyplace in the country, much less the world, if they were relying on humans to do it? It was 1997 and the Internet—with weather as in all things—demanded a new approach.

**Neilley wanted to build something that didn't need humans—or at least needed** them less. To do that, he would tap into the raw output of the weather models. Meteorologists had been using weather models since the 1980s, but most knew better than to trust them. Models were treated as "guidance"—advice one may or may not want to take. Like the early digital cameras of the era, the models' spatial resolution was relatively coarse. That made their predictions wonky in some problematic places—like populated cities on the coast. If the closest model grid point to New York City was half a dozen miles out into the Atlantic, the temperature it spat out for the city was almost always going to be wrong. Neilley's plan was to "take the models and make them sing," as he put it, mixing their outputs with historic temperature patterns, in the places where we actually live. It wasn't the kind of physics work the scientists at ECMWF were doing. Instead it relied on statistical postprocessing of the model outputs. Most days, that would be good enough; for extreme events, the humans would still be around. The scheme got him a job at the Weather Channel, which later became the Weather Company, and started him on the two-decades-and-counting challenge of constantly improving the results.

For a long time, the system worked like a funnel. Into the wide end poured a range of inputs: real-time observations, weather radar data, and the outputs of multiple weather models. The Weather Channel's human forecasters sat at the narrow end. They would take the output as a "first guess" and then adjust it based on their experience of each model's past performance and their own understanding of the weather in a given place. "The human always controlled the publish button," Neilley said. "Content would not go out the door until the human decided that it should." The system allowed for more frequent and more efficient forecasts than humans alone could create, but the humans were still a safeguard—and a bottleneck. The forecasts could be updated only so often, and—geographically speaking—there could be only so many of them.

That became a problem when smartphones arrived. The forecast had always come, in some ways, from a human. In the nineteenth century, the forecast (such as it was) appeared in the morning or evening newspaper. When my parents were kids in the 1950s, they heard it on the radio, and it wasn't yet very good. When I was a kid, I waited for it to come on morning television, delivered, I hoped, by the goofy Willard Scott, ahead of his birthday wishes to centenarians. But now it came from an app, keyed to my location, and I was looking at it several times a day. The switch to mobile devices had us checking the forecast more often from more places and seeking more precision in time as well as place.

Neilley thought the Weather Company's system should meet those new expectations. The models were ready to take on more of the load. They had gotten more accurate, more days in advance. Their output had gone beyond mere "guidance," and it was often indistinguishable from the work of the human meteorologists. They were getting more spatially precise, with higher resolutions. And there were more of them to consider, with the addition of new models that better accounted for the chaos of the atmosphere, which further improved the forecasts. Neilley and his team conceived of a new "Forecasts on Demand Engine" that whenever a user made a request—like when I opened my app—would automatically take a fresh dip in the vast pool of the Weather Company's data and pull out the forecast for that time and location. The system would still make adjustments to model outputs, like sanding off the extremes of highs and lows, but the models would "be doing 90 percent of the work," Neilley said. "We were on their coattails, no doubt about that."

At one point, the Weather Company system was incorporating 162 different model inputs. The vast majority of those were slight variations of the European model, known as the "ensemble." Looked at all together, you could hone in on the most likely weather. But there was so much to look at. The funnel had stretched to a fire hose. The models had improved to the point that the humans were adding less to the process. "There's no way a human is going to do an analysis of 162 inputs," Neilley said. He asked his team to stop fiddling with the temperature. "Look," he told them, "when you modify the temperature forecast, you're making it worse as likely as you're making it better. It's just not a good use of your time!" The human forecasters mainly worked out of the Weather Channel's headquarters in Atlanta, where they presided over a kind of logjam, with the flow of forecasts backing up before they could approve them, making them more outdated than they needed to be. To free it up, Neilley realized that he would have to take the humans out of the loop.

Yet the humans weren't entirely useless. While the models were technically precise, there were still moments that required a greater flair for nuance. (The computer, for example, had a hard time deciding between "showers" or "storms" in its plain-language forecast.) The challenge was to ensure that the system preserved "the wisdom of the forecaster," as Neilley explained this human touch. The solution was to move the forecasters from the end of the loop, where they had been a bottleneck, and put them "over the loop," so that things could run without them. "Before, they had to wait until the model came out, and then they did their thing and they posted it," Neilley said. "Now the forecast is going out whether or not they touch it." But if they needed to touch it, they still could, altering it like a photographer applying a filter. The humans would be just another input into the system.

In July 2015, without any announcement or fanfare, Neilley turned it on: From that day forward, the Weather Company's forecasting system would no longer depend on humans to share its data with the world.

The Weather Company's technical office, where Neilley works, is located on the second floor of a white and glass building in a tidy office park in Andover, Massachusetts, thirty miles north of Boston. In a conference room called Tsunami, Neilley and a software engineer named Jim Lidrbauch agreed to demonstrate how the new system worked.

Lidrbauch plugged his laptop into a giant monitor at the head of the table and fired up the program they call HOTL, for "Humans Over the Loop." It was half Google Earth, half time machine. A red slider at the bottom of the screen let meteorologists fast-forward the weather like scrubbing through a movie. As Lidrbauch moved it around, the map became overlaid with polygons, as if a five-year-old had gotten loose with a drawing tool. Each represented a change made by a human—the last vestige of active involvement on the part of the meteorologists. Anytime someone, anywhere in the world, requested a forecast for a location inside the polygon, the change would be applied to the output automatically. One filter specified "no ice," telling the forecasting engine to output either rain or snow, but not sleet, for that area (regardless of what the model said). Another blob, over Georgia, was the work of someone in Atlanta. It instructed the system to increase the cloud cover by 5 percent and decrease the temperature by one degree. "A tweak," Lidrbauch said.

Neilley, sitting at the far end of the table, squinted over the top of his laptop. "We have an active filter that says decrease the temperature? Where is that applied to? What period of time does that cover?"

"This afternoon," Lidrbauch said. He clicked around to see who made the change: a Weather Company meteorologist in Atlanta named Juan. Had Juan looked out the window and not liked the way the system agreed with the sky? Was he bored?

Neilley sighed. "Subtracting one degree Fahrenheit from Atlanta for the rest of the afternoon: Was that a good use of time? Humans by nature will fill their time with work."

The Weather Company's computerized system generates weather forecasts on demand, some 26 billion times a day, all over the world. And most of them go out without human intervention. This is more than just a change in an app (even a big one). It represents a much broader change in meteorology: The forecasters were no longer forecasting. Though Neilley insisted on calling the shift to over-the-loop an "evolution," not a "revolution," it was remarkable nonetheless. Neilley had led his team across a Rubicon. A century after it was first considered that the weather could be calculated; sixty years after the first computer weather model; and thirty years into their widespread usefulness, the modeling system had reached a true point of maturity, where most days it could power forecasts as well as a human. From here on out, it would be models all the way down. The machines were in charge—and they were busy.

The Weather Channel finally purchased Weather Underground in 2012, right before swapping out "Channel" for "Company" in its name. As when Apple dropped "Computer," the move reflected broader changes in the firm's business, and in meteorology as a whole. It wasn't about television anymore; they were an information company now. That became even clearer in 2016 when IBM purchased the Weather Company (but not the Weather Channel, which operates independently while drawing on the Weather Company's data).

The reach of the Weather Company's "forecasting engine" was breathtaking. It serves not only their own website and app, as well as Weather Underground's, but also supplies the forecasts viewed on Google, Apple, Yahoo, Facebook and countless other websites and television stations around the world. They also have professional forecasting services, for clients like airlines and power companies, and an active business looking for new uses of their data.

But for most of us, it's more personal than that. For a while when my daughter was little, she would ask the same question at bedtime, just before rolling toward the wall and into a deep sleep: "What's it going to be tomorrow?" The question was innocent, of course: a prompt for dreams, a thought to carry her off to sleep. She meant it practically. Was the next day a school day or the weekend? Did we have anything planned—a playdate, an obligation, some bigger adventure? Four-year-olds have a poor sense of time, and so the grind of the calendar, the fast and slow rhythm of days, weeks, seasons, lives, didn't occur to her. All her days were "tomorrows." The question also seemed existential. She wanted some reassurance that tomorrow would be there when she opened her eyes. That the sun always rose after the dark.

Her question, "What's it going to be tomorrow?" wasn't meteorological, but I often found myself wishing it were. It was the easiest question to answer. I could find out in three seconds flat. I could open up one of half a dozen weather apps I kept on my phone and see the symbols: the little clouds and suns, the occasional snowflake and parallel lines, indicating wind. If it were Weather Underground (which it usually is), I would be using the forecasting engine that Neilley created. These things change often, but in its current iteration, a red line undulates diurnally against a faint grid, its peaks and valleys registering the afternoon highs and the nighttime lows for the next ten days. The most interesting part is how it changes. Over hours and days the shape of that red line subtly shifts like a snake in the grass. If I check in the morning, the high for three days out might be 45 degrees. When I check again in the afternoon, it might show 43—except that forecast is no longer three days out but two and a half. The forecast is never static. Its future is always getting closer (that's how time works). But the forecast itself is also changing, honing in on the most likely weather with each run of the models, right up to the moment when the future becomes the present. The forecast is less a rhythm than a flow. It is drawn from a sequence of processes, calculations and observations, cascading back through the system, stretching around the world, high into the atmosphere and deep into space. It is the final point of a long supply chain of data, like the tip of a giant hidden iceberg. By the time all of that gets onto the screen in my hand, it has coalesced, becoming merely "the weather," the handsome face of a complex and sprawling machine. With its help, I can know (or at least have a pretty good sense) what the temperature is going to be at every hour. From experience, I can imagine what that will feel like—how 15 degrees cuts in the throat, and 92 feels like a weight.

The little pictograms on my screen aren't guaranteed to be correct, but they are usually pretty close. How we face our tomorrows, how we strut and fret our hour upon the stage, is an open question—the one we spend our lives learning to answer. But what's the weather tomorrow? The weather machine makes that one easy.

What we do with it is harder.

# 10

# The Good Forecast

My conversation with Tim Palmer was a bit of an afterthought. While I was at the European Centre in Reading people kept mentioning his name. He was a kind of éminence grise—an occasional visitor, who'd moved on to a professorship at Oxford, not far away. His claim to fame was a funny idea, given the place's obsession with better forecasts: He had argued that weather predictions are not perfect, and they never will be. His suggestion, instead, was for "more accuracy with less precision," an oxymoron that he characteristically enjoys.

"This happened to me a few years ago," he said, as we sat on two low chairs in a corner of the Weather Room. He wore a polo shirt beneath a stretched-out blue sweater and had a big tuft of curly gray hair. On the coffee table in front of us was a brochure for a new NASA satellite, like a movie magazine at the dentist's. "Somebody phoned up, and it was their wedding anniversary or something, and they were having a big party," he said. "They had to know by Monday if they should get this marquee on a Saturday." So they needed to know: "Was it going to rain? Or not?"

It was the sort of question the weather modelers in this building dream of, one would think. Their objective was a perfect model that was right every time. But that was the wrong ambition, Palmer argued. "Life's not necessarily as simple as that. What you have to decide is how important is it that the people who are coming to your party don't get wet? I would suggest that if the queen was coming, for example, and I told you there was a 10 percent chance of rain, you might want to go for the marquee because if the queen got dumped on and you were looking for some big knighthood, this would not be a good thing—so spend all that money just on a 10 percent chance. But on the other hand, if it was just your mates from the pub, you might not bother getting a marquee unless it was about 80 percent probability." Palmer squinted his eyes at me. "The point here is knowing that uncertainty, knowing that probability, allows you to make a much more informed decision than if I just said, 'Yes. It's going to rain.' Or, 'No, it's not going to rain.'"

**What is a weather forecast good for? What is a good weather forecast? What is a** weather forecaster good for? The achievements of the weather machine in its current incarnation have pushed meteorology into a period of transition. Peter Neilley may insist on calling the shift at the Weather Company an "evolution," not a "revolution," but it's remarkable nonetheless: The forecasters are doing less forecasting. Instead, the weather models, and the systems that sit between them and us, like the Weather Company's forecasting engine, are doing more of the work.

If the Weather Company was at the forefront of this transformation, it is no longer alone. For example, in 2017, the National Weather Service still had a corps of twenty-five hundred meteorologists tasked, in part, with hand publishing what the Weather Company managed with a staff of thirteen. As it actively works toward a more automated system, the Weather Service has shifted the staff's priorities. "We're fundamentally changing where our job actually ends," Louis Uccellini, the director of the National Weather Service, told me. Most notably, that means spending more time explaining to emergency managers and public-works officials the likelihood of a weather event and the severity of its impacts. In the interim, that has made for more work. In the future, it could well be the only work.

A new generation of television meteorologists has already come to terms with this. Ryan Hanrahan, an on-camera weatherman at NBC Connecticut, has extensive scientific training, but he's using it less and less for the kind of day-to-day forecasts that would have kept him busy even a few years ago. (Big storms are a different story.) "There's no question we're going to move to a role that's more about communication than actually figuring out if, three days from now, it's going to be 66 or 68 degrees," he said. Yet the recognition of this paradigm shift in the role of meteorologists has been unevenly distributed. "I think some are in denial that computers can do as good a job as you can," Hanrahan added. That may be a reasonable response if you have spent decades looking at the outputs of weather models and could tell that they were no good. But as the models have unequivocally improved, the bar has been raised; it is more often harder, if not impossible, to outsmart them.

Paradoxically, it's precisely these improvements in the automated forecasts that have led to a new emphasis on communication. When the forecast was wrong half the time, decisions were harder to make. Flights were canceled later, schools closed only after snow had already fallen. Today's forecasts are good enough to be actionable, often several days in advance. Which raises a new challenge: If the weather forecast is nearly perfect, what can you do with it? How do you learn to make decisions using it? In the past, meteorology had been slow to address this reality. "It was always an afterthought for our science, it was somebody else's problem," Neilley explained. "Our science for a long time said, 'We're just going to focus on accuracy and then when we reach utopia in accuracy, society will be in good hands.' But we have realized that that is not entirely true." Their work has expanded. It now includes "the entire value chain, from the production in the forecast starting in the models all the way to an effective decision by an individual," Neilley said.

At the National Weather Service, the forecasters' union has been understandably suspicious that its members will be replaced by robots, but Uccellini frames it as a return to a basic sense of purpose. "If you look at the Weather Service mission, the first part is to 'produce and deliver observations, forecasts and warnings of weather, water and climate,'" he said. But the second part is to "save lives and property and enhance the national economy." Uccellini likes to quote a line from Allan Murphy, a professor of meteorology at Oregon State University known for his eloquence and clear thinking, who died in 1997. "Forecasts possess no intrinsic value," Murphy wrote. "They acquire value through their ability to influence the decisions made by users of the forecasts." It matters what we do with the forecasts, not merely that they correctly predict the weather.

Why check the weather? We rely on the forecast, to varying degrees and with shifting stakes, for its ability to tell us what the weather will do next and how much it will matter to us. "The sole purpose of making weather forecasts is to aid decision making," Tim Palmer has written. "As a daily commuter, should I take my umbrella to work? As a regional governor, should I order the evacuation of a coastal city ahead of some possible hurricane? As an aid worker, should I prepare for relief measures ahead of an ongoing drought?" The remarkable capability to answer those questions comes from the satellite and the supercomputer, the observation station and physicist. It comes from the hundreds of scientists at the European Centre focused on creating a better simulation of the atmosphere. But it is not perfect—and even its near perfection is limited in time (three days, four days, maybe five days . . .). As the weather machine keeps improving, the challenge will be to evolve to meet the needs of its own success. Just in time, one hopes, to meet the needs of a desperate planet.

# PART IV

# PRESERVATION

# 11

# The Weather Diplomats

On the Monday morning of July 1, 1776, Thomas Jefferson arrived at the Pennsylvania State House in Philadelphia. For the previous two months, he and his fellow delegates to the Second Continental Congress had been debating the future of "these united states," as Jefferson called them, wishfully. His draft of their "declaration of independency" was complete, and all that was left was the delegates' commitment and their signatures. It would be a busy, consequential week. All of which makes it surprising, strange even, that at nine o'clock Jefferson took out a thermometer and measured the temperature. It was 81½ degrees.

Why, at that fraught moment, would Jefferson be thinking about the weather? It had to have been premeditated. The thermometers of the time were big, as tall as candlesticks, their glass tubes protected from breakage by heavy wooden carrying boxes. Jefferson's was brand new. He'd bought it that week at Sparhawk's, a bookstore a few blocks from the statehouse. It was expensive: three pounds and fifteen shillings, a few hundred dollars today. We know Jefferson loved buying gadgets and also designing them. Already in Philadelphia he had commissioned a cabinetmaker to build a revolving chair and a portable writing box (on which he wrote the Declaration of Independence). Like any good gadget freak, Jefferson was likely eager to play with his new device. Thermometers were rare, with none made in America at the time. I can imagine the other delegates gathered around him at the breaks to admire it. Franklin would have wanted a particularly close look. John Adams, ever the Yankee, might have scorned the expense. Jefferson recorded the temperature a second time on July 1, at 7 p.m., when it was 82 degrees. The following two days, he managed to take three observations. On July 4—when one would have expected him to be preoccupied—Jefferson found time to take four separate observations. Lore has it to be an uncomfortably hot day, but Jefferson's records show that it was pleasant by the standards of Philadelphia in July: 68 degrees at 6 a.m., 72¼ degrees at 9 a.m., 76 degrees at 1 p.m. and 73½ degrees at 9 p.m.

Jefferson's weather project began with the birth of the nation and continued with the life of the nation for fifty years. In 1788, he designed a thermometer "intended to be hung on the outside of a glass window, with the face of the plate next to the window so that it may be seen without opening the window," as he instructed the instrument maker. At Monticello, his estate in Virginia, he installed a weathervane with an indicator fixed into the ceiling, so that it could be read indoors. In 1790, when Jefferson became the first secretary of state, he moved to New York City, and began looking for ways to expand his weather observations. He had two problems, the first familiar still: "I find it difficult to procure a tolerable house here," he wrote to his daughter from the city. But the bigger problem with his real estate search was that it delayed his comparison of the weather in New York and Virginia. "As soon as I get into the house I have hired I should like to compare the two climates by contemporary observations," Jefferson wrote to his son-in-law Thomas Mann Randolph, who he hoped would help him with the project. Germany and France had recently established shared weather observation networks, but there is no indication that Jefferson knew anything about them, as closely as he followed European scientific developments. Jefferson worked through the technical challenges on his own. "My method is to make two observations a day, the one as early as possible in the morning, the other from three to four o'clock, because I have found four o'clock the hottest and day light the coldest point of the twenty-four hours," he wrote. "I state them in an ivory pocket book in the following form, and copy them out once a week. Otherwise the falling weather would escape notation." Anticipating that the twenty-two-year-old husband of his daughter might roll his eyes, Jefferson reiterated his purpose. "I observe these things to you, because in order that our observations may present a full comparison of the two climates, they should be kept on the same plan."

By 1797, Jefferson's weather ambitions had grown again. In a letter to the French philosopher and politician Constantin François de Chassebœuf, he speculated about a broader observation system, one stretching across the new country. "As I had then an extensive acquaintance over this State, I meant to have engaged some person in every county of it, giving them each a thermometer to observe that, and the winds twice a day for one year, to wit at sunrise and at 4 pm (the coldest and warmest point of the twenty-four hours), and to communicate their observations to me at the end of the year," he wrote.

Jefferson wasn't thinking of weather prediction. He had no reasonable expectation of communicating his observations faster than storms moved. (He would die a decade before the first telegraph experiments.) Yet Jefferson intuited a broader and more lasting point that has carried through until today: that the weather ties the world together. His observations were not only a scientific project but a political one. All weather measurements are. His thermometer-wielding deputies could use their instruments as the screws of unity. It was a classic Jeffersonian insight, combining the political and the natural, the individual and collective. He recognized that we live on a planet carved up by borders but encased in a borderless atmosphere.

**We still do, of course. For the past hundred and fifty years, the task of navigating** the tension between the earth's fluid atmosphere and its fixed political borders has fallen to the weather diplomats of the World Meteorological Organization and its predecessor, the International Meteorological Organization. It was to them that John F. Kennedy sent his call for a global observation system, and they have, in the decades since, been committed to its development and continuity. But the challenge—then, today and all the way back to Jefferson's time—is keeping the coalition together. A global weather machine is not self-evident. What makes it keep on? Especially at a moment when changes in the earth's climate and new extremes of weather make it so crucial?

Every four years, weather diplomats gather in a conference center in Geneva for "Congress," as the meeting is known. Its agenda is long and varied, dragging on for nearly a month, but the crux of the effort hews close to the long-held dream: the international exchange of weather data. Ever since the fourth congress in 1963, that international exchange of weather data has specifically meant the furtherance of the World Weather Watch. But by the time of the seventeenth congress, in 2015, attended by delegates from 191 states and territories, the terms of that weather exchange had become more fraught, making it a precarious moment in the life of this global data system.

Half a century ago, weather observations flowed usefully in all directions, a matrix encompassing the world. Even early satellite data—while produced exclusively by the superpowers—was distributed everywhere. Beginning in 1966, the American satellite known as ESSA-2 automatically transmitted real-time pictures to any country with the equipment to receive it. This imagery, along with the systematic exchange of more traditional observations, could immediately be put to use in the creation of local forecasts by local forecasters. That was the goal of the endeavor. "We don't take observations for the sake of observations," as Dr. Sue Barrell, deputy director of Australia's Bureau of Meteorology, put it. "We take them because they underpin the services for our community."

But increasingly, the core of those services—meaning weather forecasts, warnings and other kinds of analyses—is derived not from the work of local forecasters but from the outputs of global weather models. With that shift, the way data is exchanged internationally has changed as well. If once a diagram of its movement resembled a spiderweb, with lines connecting each place to every other place, the current flow looks more like the route map of an international airline, with a few capitals sending out far more information than they are receiving. What was once a many-to-many structure of data exchange is now a much smaller club. As the balance of forecasting power shifts toward the models, it also becomes more consolidated. The most useful data produced by weather satellites is often the most complex. In particular, the quantitative data produced by the instruments on the polar orbiters, like EUMETSAT's Metop satellites, is a crucial input to the weather models, but it is of minimal use to the smaller weather offices that don't do their own modeling. As in so many realms, the world is becoming split, and the gap is growing.

Almost every nation has some form of weather service, and that's been the case for a century. In the parlance of the WMO, they are NMHSs, or National Meteorological and Hydrological Services. From an international perspective, the core task of NMHSs is to make weather observations and share them through the carefully structured networks of the World Weather Watch. But the complexity and breadth of the quantitative satellite data has struck a new hierarchy of nations and their weather services, from the large centers with the expertise and budgets to run their own weather models (if not their own satellites) down to the smaller nations that depend on them. The quick, broad and often free dissemination of weather forecasts over the Internet obscures the reality that the world's forecasts are increasingly dependent on a narrowly constructed system. This mirrors broader changes in our use of technology in disconcerting ways. When the Weather Company sells its global forecasts to Facebook, and Facebook is a nation's major source of news, where does that leave the nation's weather service?

Similar forces of technological change are at work against the observation networks as well. There is the possibility that billions of tiny temperature and barometric sensors—in smartphones, home devices, attached to buildings, buses or airliners—could meaningfully compete with the relatively few and carefully constructed weather stations of the Regional Basic Synoptic Network. That isn't yet the case, and there are plenty of technological hurdles in the way. But there is a major diplomatic one as well: Who would own the data? Government weather services have a hundred-and-fifty-year history of sharing their data and giving their services away for free. But if observations are being made by private networks and aggregated by the Googles, IBMs or Amazons of the world, that openness can no longer be assumed. The weather machine is based on an idea of international cooperation that has become outmoded. Many of its interdependent parts were based on colonial structures. Now multinational technology corporations are poised to create a new structure of data ownership and exchange. How will the weather machine adapt to a world networked in new ways?

It must, because we need weather forecasts like never before. With each passing congress of the WMO, the effects of earth's changing climate have crept upward on the agenda. By the seventeenth congress, they had brought a new focus and determination to each delegation's presence. As Ban Ki-moon, the United Nations secretary-general at the time, put it (in a videotaped welcome message projected on a screen above a dais festooned with a banner of clouds): "As the global thermostat continues to rise, meteorological services are more important than ever."

New weather extremes demand new efforts to smooth over the differences among nations and new stakes for the group's understanding for how weather ties us all together. "There is a deeply ingrained sense of international cooperation in the meteorological community that flows from the nature of the global atmosphere," John Zillman, a former director of Australia's Bureau of Meteorology, told me. And yet all that peace, love and understanding is balanced by the darker realization that the global atmosphere will be the source of future upheaval.

The main meeting hall of the Palais des Congrès, a Brutalist building in the heart of Geneva's diplomatic district, is a room the size of a concert hall, arranged with long lines of desks. In there, at least, everything is perfectly equitable. No matter the size of the nation, each delegation is assigned four swivel chairs upholstered in maroon leather. In front of them are set plastic nameplates printed in proper Swiss Helvetica. With the delegations arranged alphabetically, in French, there are strange bedfellows, like les Etats-Unis between Estonie and Ethiopie and not far from Iran. Seeing meteorologists from all over the world gathered together in a common cause, what seems strangest of all is how plain it made the fact that this was the only world we had—that this was our only planet, our only home.

Each national delegation to congress was led by its "permanent representative"—only ever called "the PR"—who was nearly always the head of each nation's weather service. The glaring exception to that custom was the United States, even before the Trump-era retreat from the international community. The

National Weather Service sent not its director but its deputy director, a gesture universally understood as a slight to the weather community as a whole. To compensate, or maybe rather to reiterate the arrogance of the gesture, the United States contributes 20 percent of the WMO's budget, double the next largest member (Japan) and three times the amount from G7 states like France and Germany. (The formula is determined in parallel with funding to the United Nations as a whole.)

Along the edges of the Palais des Congrès, the richest weather nations set themselves up in offices rented for the duration of the meeting. When I was there the United Kingdom had taped a Union Jack in its hallway window. The United States had pushed the desks in its office together into a single big table, which was piled with chips and cookies. With the exception of the Saudis and Emiratis in white gowns, everyone wore a dark suit, men and women both. But what the crowd lacked in the originality of their fashion, they made up for in the variety and enthusiasm of their greetings: handshakes and bows, two kisses and three, backslaps and stiff lips. The sense of an international community was palpable and thrilling.

I found an empty seat in the back row of the hall. In front of me was the representative from the Holy See—who had "observer" status, since the Vatican does not have its own weather bureau. He'd set out a mouse pad next to his laptop, a tiny fringed oriental rug. Congress was a marathon: three weeks long, six days a week, nine to five thirty on weekdays, half days on Saturdays. If a delegation wanted to speak, they pressed a button on a small screen in front of them and a light turned green. The interpreters sat in glass booths at the top of the tiered room. Regardless of the language being spoken, everyone wore their translation headphones, all the time, which kept a hush over the hall. If once the focus in Geneva was primarily on how to coordinate observations, many delegates at the seventeenth congress were visibly concerned with making this new world order work and keeping the weather machine humming. Like the United Nations itself, their alliances and desires were based less on geographic proximity and more on economic dependence. Smaller nations, particularly those facing new climate extremes, needed more than ever the satellites and weather models of larger and richer nations. The meteorological director from Madagascar painted a stark portrait of the "natural cataclysms" that had befallen his country. "Despite the cyclones and droughts, the government is sparing no effort—*with the valuable assistance of the international community*," he emphasized, in French, "to mitigate the suffering and pain felt by the local population, in a context generally characterized by loss of life, many hectares of flooded crops and damaged infrastructure." The minister of environment for Cabo Verde pointed out that it was African Unity Day and called for a "more focused attention" on "small island developing states." Some of the delegates naturally commanded the room, while others struggled to press the button on their translation consoles with visibly shaking hands. The Namibian representative used his time for a different point: "Since this is the first time that Namibia is taking the floor, please allow us to express our appreciation to Switzerland for their warm welcome."

Before lunch one day, a staffer from the WMO secretariat—the office in Geneva that implements the decisions of the congress—read a tweet directed at us from the Italian astronaut Samantha Cristoforetti, who was at that moment orbiting the earth on the International Space Station. "Our atmosphere is captivating and powerful, understanding it a challenge." In response, all of us in our dark suits and translation headphones waved at a tiny camera on the big stage, which snapped our photo to be tweeted back to space. What at first felt like a silly social media moment then seemed powerful: this direct connection with a scientist looking back at the earth from space, this link to the ur-principle of the weather machine.

Weather diplomacy may be nuanced, but its societal benefits are tangible, to every country on earth. Weather services reduce the human impacts of natural disasters, make transportation safer and more economical, and help use natural resources more sustainably. WMO estimates put the economic value of weather services in excess of $100 billion annually, while the cost of providing them is a tenth that. The whole thing can be thought of as "the most successful international system yet devised for sustained global cooperation for the common good in science or in any other field," as Zillman, the Australian director, has put it. Collectively, the people gathered at the Palais des Congrès represented "one of the world's most widely used and highly valued public goods"—a statement with which it's difficult to argue, even if it's rarely heralded. The meteorologists, scientists, ministers, bureaucrats and their assistants—representatives from almost everywhere on earth—were eager to ensure its continuation. But the tweets, ceremony and gentle discussion couldn't obscure the challenge the system faced.

The congress had previously wrestled with the question of how to deal with the changing flows of weather data twenty years earlier. In particular, the 1990s vogue for the privatization of government services had pushed some weather bureaus to sell the meteorological data that, for more than a century, they had freely exchanged. After a contentious debate, the WMO's members drafted what became known as Resolution 40, a fundamental agreement that required members to provide "essential data" on "a free and unrestricted basis." In a decade that saw a realignment of global alliances following the collapse of the Soviet Union, Resolution 40 was an occasion to reaffirm "worldwide cooperation in the establishment of observing networks" as a fundamental obligation of WMO membership.

But in the years since, what constituted "essential" had evolved along with the different ways the weather was being observed. The discussion came to a head during the congress session listed on the agenda as "open data policies and their impact on WMO stakeholders." There were three interrelated concerns on the table: open data, big data and crowd sourcing. The prospect of massive amounts of new observations from tiny sensors, aggregated by big technology corporations, was particularly threatening. The more technically minded among the crowd were concerned with quality. By tradition, the WMO demanded high standards of observations to maintain their consistency and accuracy: Wind measurements had to be taken at a certain height, thermometers housed in a certain way and stations properly sited to avoid extremes of sun or wind. If the Global Observing System was going to open the floodgates to crowd-sourced data, aggregated by private companies, then new policies would have to be established to figure out how to manage and share it. Inevitably, the observing system as it existed—heavily structured, carefully assembled and codified, global, shared, primarily government-run—would be threatened in a world filled with billions of new sensors, privately operated satellites and new, Google-like capabilities for aggregating data. Ironically, the expense and complexity of the existing satellite system offered some ballast to this possibility: Its data still dominated the models. The complexity of the satellites and their instruments meant that they changed slowly and deliberately, offering some reassurance that nothing would happen fast. And yet the tension at the WMO over how to deal with these new types of crowd-sourced and private data fundamentally reflected the broader technological changes outside of meteorology.

How was an organization organized by nation-states, committed to open data and borne of a global view supposed to work in a world in which the default mode of information was to live on private platforms and travel across private networks? As with so much else this decade, the technological changes were fundamentally anathema to the basic charter of the WMO and the world order it reflected.

**The question preoccupied David Grimes, the WMO president. The WMO has two** leaders. The secretary-general, who manages the WMO secretariat in Geneva, is a full-time paid position, with the job of implementing the policies of the congress. In contrast, Grimes's full-time job is head of the Meteorological Service of Canada, but as president of the WMO he effectively led the proceedings. I'd admired how he managed the discussion in the hall.

"Excellencies, ladies and gentlemen, colleagues," he would call, occasionally banging on his maple gavel. When there was any disagreement or frustration from the floor, Grimes would never address it head-on but instead move to the next country in the queue on the console in front of him, like a patient teacher. "Is that okay?" he would ask, hopefully, of a counterpoint. "Acceptable? Okay." Occasionally, he would defer, knowing that things either had to keep moving or there were other battles to fight. "I don't know the appropriateness of having this in the resolution," he'd gently push, on a matter of slim detail. "It's not 'if it is or isn't,'" he said once. "It's 'does it belong here or does it belong there?'" Diplomacy was subtle. Every couple of hours, he'd lead the hall in a "seventh inning stretch," a provincial suggestion that always got a laugh.

During a break, I sat down with him in an office upstairs in the Palais des Congrès that he had commandeered for the duration of the session. His concern for the preservation of the existing data regime was clear. "Technology is moving so fast that we don't have systems that are scalable enough to manage it," he said. But implicit in that was a big, and hopeful, "yet." In the early days, the WMO's mission was all about consolidating observations from particular places on earth. After the first weather satellites were launched, that data wasn't merely sent from sovereign territories but was instead a picture of the earth from above, using the tools of a particular nation—the wealthy ones. More startlingly, as the satellite data evolved to include quantitative data, it required integration into the global weather models to be useful. "Say you've got something about water vapor, and water content and ozone and a bunch of other stuff," Grimes said, plucking an example. Their values were collected by the current generation of instruments—at a global scale—but their usefulness was much more limited. It depended on a very high level of computational analysis. "The reality is that if you're a small met service, there's no way that you can integrate any of those pieces," Grimes said. "But if you run a modeling center, you can." The complexity of today's weather data was unassailable. So how, in the next era of data and modeling, would the WMO evolve?

"What are the risks and opportunities, and what are the ways forward?" Grimes asked. In a session of congress devoted to these new types of data, he'd challenged the working committee to think broadly about the question—to potentially redefine the next generation of international exchange, for a new epoch of data. "My call to them was, just don't write a paper that I could copy off Google where someone wrote about big data," he said. "That's not going to be helpful." There had to be a concerted effort to redefine the system—a diplomatic effort as much as a technological one. It seemed a modest ambition, but his point was that they had to stay together. The diplomatic process could go on slowly, meeting by meeting, resolution by resolution. "But you have to make everyone feel comfortable when they take a step toward each other," he said. "You got to make them all feel like they're winning something."


**That evening, we were all winning. Diplomacy meant parties**—or, rather, "receptions." The largest of them were hosted by the nations with a candidate in the race for secretary-general. The director of the

Finnish Meteorological Institute, Petteri Taalas, was the leading contender for the post, up against the current deputy secretary-general, Jeremiah Lengoasa from South Africa. During the apartheid years, South Africa had been suspended from the WMO, and in a lunchtime reception at the Palais des Congrès, the South African delegation presented Lengoasa's experience and what his appointment would mean to the nation. It gave a poignancy to his candidacy, as someone who had grown up, as he put it, "on the dusty streets of Soweto." But the buzz in the halls was that Lengoasa's candidacy was a long shot.

As if to emphasize the point, in support of Taalas's candidacy the Finns had chartered a yacht for a dinner cruise on Lake Geneva, and everyone was invited. The receiving line stretched up to the quay, and Taalas, facing into the evening sun, shook everyone's hand as we boarded the boat. Here was the entire diplomatic weather community, jovial and chatty, eating meatballs and potatoes on heavy banquet service plates.

This wasn't exactly what I had expected to find when I set out to understand the weather machine, but it made a certain amount of sense. This global infrastructure, like all infrastructure, had been built in the image of its creators. It was a product of the international order, and the rituals of the congress were there to maintain the status quo of data exchange, as they had for three generations. In narrow terms, longer-term weather forecasts require the broader collection of observations. There is no four-day weather forecast without four-hemisphere observations. For a small nation—or even a large one operating satellites—it is easy to point at what they are getting in return for their contributions to the global observing system. That is crucial to the whole endeavor: You can't look at the weather alone, within your borders. Or, rather, you could, but you won't get very far; the weather over the horizon—in space as well as time—will come as a surprise. The weather machine has to be a global system, and it won't work any other way. At its heart is an equilibrium between the things nations do for themselves and the things they contribute to systems that supersede their borders. We are many countries, on one planet.

And yet, the technological winds are blowing against us. The most important weather observations are increasingly collected by the narrow tier of countries that operate satellites. And the most important forecasts are produced by the equally slim group of countries (or groups of countries) that operate weather models. How long can the current system of data exchange among nations hold? How soon might it be supplanted by global technology corporations—themselves often acting like nations? The weather machine is a last bastion of international cooperation. It produces some of the only news that isn't corrupted by commerce, by advertising, by bias or fake-ness. It is one of the technological wonders of the world. At the beginning of an era when the planet will be wracked by storms, droughts and floods that will threaten if not shred the global order, the existence of the weather machine is some consolation.

A week later, Taalas won the election for secretary-general, not replacing Grimes as president but running the permanent bureaucracy of the WMO. The ship of weather diplomacy kept on its course.

# ACKNOWLEDGMENTS

This book benefited from unusually good circumstances: enthusiastic and patient publishers, an attentive agent and unfaltering support from my family. I am grateful for the privilege of time and space to write, and I hope these pages honor the trust and opportunity each extended. At Ecco, Daniel Halpern, Miriam Parker, Dominique Lear, Emma Janaskie, Denise Oswald and Hilary Redmon were my lighthouse keepers, keeping this project off the rocks and brightly guiding it to completion. For nearly a decade, my writing and thinking has benefited from the subtle and precise suggestions of Will Hammond at the Bodley Head in London. I am grateful for the continuing support of Jim Gifford at HarperCollins in Toronto and Britta Egetemeier and Julia Hoffmann at Penguin Verlag in Munich. My agent, Zoë Pagnamenta, has the uncanny knack for doing just the right thing at just the right moment, while Alison Lewis is equally precise and unflagging. Robert Pincus, my neighborhood cloud scientist, was an indispensable tutor and mentor, pointing out (and opening) many doors. My parents, Diane and Ron Blum, could teach a course on when to ask about a book and when not to. Phoebe's and Micah's love of books and reading has been the best daily reminder of why I write. And this book exists only because of Davina's clarity of thought and love.

# NOTES

## PROLOGUE

2 **doctors carried twenty-one infants:** Michael Espiritu, Uday Patil, Hannaise Cruz, Arpit Gupta, Heideh Matterson, Yang Kim, Martha Caprio, and Pradeep Mally, "Evacuation of a Neonatal Intensive Care Unit in a Disaster: Lessons from Hurricane Sandy," *Pediatrics* 134.6 (2014). http://pediatrics.aappublications.org/content/134/6/e1662.

2 **Across the region:** National Oceanic and Atmospheric Administration (NOAA), "Service Assessment: Hurricane/Post-Tropical Cyclone Sandy, October 22–29, 2012," U.S. Department of Commerce, National Oceanic and Atmospheric Administration, National Weather Service (May 2013), https://www.weather.gov/media/publications/assessments/Sandy13.pdf.

4 **six-day forecast today:** P. Bauer, A. Thorpe, and G. Brunet, "The quiet revolution of numerical weather prediction," *Nature.* 525.7567 (2015): 47–55, https://doi.org/10.1038/nature14956.

## CHAPTER 1: CALCULATING THE WEATHER

Anyone curious about the history of American meteorology quickly comes to admire the comprehensive work of James Rodger Fleming of Colby College. I am indebted to his seminal research on nineteenth-century American meteorology, as well as his more recent study of Vilhelm Bjerknes and Harry Wexler (who features prominently in chapter four). In Oslo, Anton Eliassen, Yngve Nilsen and Gabriel Kielland at the Meteorological Institute were gracious and joyful in answering all my questions, with thanks to Heidi Lippestad for facilitating. Einar Sneve Martinussen and Jørn Knutsen, both of the Oslo School of Architecture and Design, were game and knowledgeable on a tour of the city's weather stations. I was lucky not only for the existence of Robert Marc Friedman's wonderful biography of Bjerknes, *Appropriating the Weather*, but also his early personal encouragement of my explorations. And I could imagine few more expert (or patient) tutors in circulation theory and the primitive equations than Adrian Simmons and Alan Thorpe of the European Centre for Medium-Range Weather Forecasts; any misunderstandings are, of course, my own.

13 **worked poorly in the rain:** James R. Fleming, *Meteorology in America, 1800–1870* (Baltimore: Johns Hopkins University Press, 1990), 143.

13 **"If I learned from Cincinnati":** Ibid.

13 **"considerable comment and wonder":** Ibid.

13 **"the globe has been practically reduced in magnitude":** James Gleick, *The Information: A History, a Theory, a Flood* (New York: Vintage Books, 2012), 148.

14 **"a widespread and interconnected affair":** Ibid., 147.

14 **"perfect systems of methodical and simultaneous observations":** John Ruskin, "Remarks on the Present State of Meteorological Science," from *Transactions of the Meteorological Society* (1839), 56–59, as quoted in Paul N. Edwards, "Meteorology as Infrastructural Globalism," *Osiris* 21 (2006): 229–50, https://doi.org/10.1086/507143.

15 **"from weather science to weather service":** Fleming, *Meteorology in America*, 141.

15 **paper disk the size of a poker chip:** Ibid., 143.

15 **"This map is not only of interest":** Mark Monmonier, *Air Apparent: How Meteorologists Learned to Map, Predict, and Dramatize Weather* (Chicago: Univ. of Chicago Press, 1999), 41.

16 **"into a single interconnected nation":** Lee Sandlin, *Storm Kings: The Un-*

*told History of America's First Tornado Chasers* (New York: Pantheon Books, 2013), 77.

16 **"in the leaden morning":** as quoted in Peter Moore, *The Weather Experiment: The Pioneers Who Sought to See the Future* (New York: Farrar, Straus and Giroux, 2015), 236.

17 **Meteorological Office soon had fifteen telegraph stations:** Monmonier, *Air Apparent*, 45.

17 **In 1864, the International Geodetic Association:** Paul N. Edwards, *A Vast Machine: Computer Models, Climate Data, and the Politics of Global Warming* (Cambridge, MA: MIT Press, 2010), 50.

17 **met in Vienna in 1873:** Ibid., 51.

17 **"distant points of the Earth's surface":** Edwards, "Meteorology as Infrastructural Globalism," 232.

18 **two observation stations per each quadrangle:** Ibid.

18 **What is the best form:** *Symons's Monthly Meteorological Magazine*, April 1873 (London: Edward Stanford).

18 **"telegraph the elements of present weather":** Lewis F. Richardson, *Weather Prediction by Numerical Process* (Cambridge: Cambridge Univ. Press, 1922), vii.

19 **"Meteorology has been handsomely supported":** Cleveland Abbe "The Needs of Meteorology" *Science* 1.7 (1895): 181–82.

20 **the most famous portrait:** 1983 painting by Rolf Groven on view at the Geophysical Institute in Bergen, https://bjerknes.uib.no/en/article/news/pioneers-modern-meteorology-and-climate-research.

20 **traveled together to Paris:** Robert M. Friedman, *Appropriating the Weather: Vilhelm Bjerknes and the Construction of Modern Meteorology* (Ithaca: Cornell Univ. Press, 1989), 12.

20 **filled with technological wonders:** K. G. Beauchamp, *Exhibiting Electricity* (London: Institution of Engineering and Technology, 1997), 163.

21 **"two oscillating spheres mounted":** *Popular Science Monthly* 21 (Popular Science Pub. Co., etc., 1882): 253–57.

21 **"I can hardly dry the apparatus":** James R. Fleming, *Inventing Atmospheric Science: Bjerknes, Rosby, Wexler, and the Foundations of Modern Meteorology* (Cambridge, MA: MIT Press, 2016), 15.

22 **returned to Paris:** Ibid., 17.

22 **"drinking beer and discussing science":** Friedman, *Appropriating the Weather*, 14.

22 offered only "exposure" instead: Ibid., 22.

23 thirty-six pigeons: Alec Wilkinson, *The Ice Balloon: S.A. Andrée and the Heroic Age of Arctic Exploration* (New York: Vintage Books, 2013), 91.

23 Forty thousand people had gathered: Ibid., 12.

23 the idea of *circulation*: Alan J. Thorpe, Hans Volkert, and Michał J. Ziemiański, "The Bjerknes' Circulation Theorem: A Historical Perspective," *Bulletin of the American Meteorological Society* 84.4 (2003): 471–80; Friedman, *Appropriating the Weather*, chapter 2, "The Turn to Atmospheric Science"; Fleming, *Inventing Atmospheric Science*, 18–21.

24 they put their heads together: Friedman, *Appropriating the Weather*, 37.

24 helped Bjerknes get his hands on: Ibid., 38.

25 "I want to solve the problem": Quoted in Friedman, *Appropriating the Weather*, 55.

26 "main task of *observational* meteorology": Vilhelm Bjerknes, "The Problem of Weather Prediction, Considered from the Viewpoints of Mechanics and Physics," *Meteorologische Zeitschrift* 18.6 (2009): 663–67, translated from German and edited by Esther Volken and Stefan Bronnimann.

26 observations consisting of seven variables: Peter Lynch, "The Origins of Computer Weather Prediction and Climate Modeling," *Journal of Computational Physics* 227.7 (2008): 3,431–444.

28 in a speech in Leipzig: Vilhelm Bjerknes, "Meteorology as an Exact Science," *Monthly Weather Review* 42 (1914): 11–14.

## CHAPTER 2: THE FORECAST FACTORIES

Lewis Fry Richardson is among the more colorful of meteorology's many characters, and it's a shame Oliver Ashford's detailed biography is so hard to find. My understanding of Bjerknes's time in Bergen was enhanced by a wonderful day spent there with Gunnar Ellingsen and Magnus Vollset, who took time off from their own work on Norway's meteorological history to shuffle through the maps of the Vervarslinga på Vestlandet with me.

31 Bjerknes received a letter: The letter is quoted in John D. Cox, *Storm Watchers: The Turbulent History of Weather Prediction from Franklin's Kite to El Niño* (New York: John Wiley, 2002), 158.

32  **"intentionally guided dreaming":** Quoted in George Dyson, *Turing's Cathedral: The Origins of the Digital Universe* (2013), 156.

32  **steampunk fantasy:** Oliver M. Ashford, *Prophet—or Professor?: The Life and Work of Lewis Fry Richardson* (Bristol: Hilger, 1985), 33.

32  **best layout for their bogs:** J. C. R. Hunt, "Lewis Fry Richardson and His Contributions to Mathematics, Meteorology, and Models of Conflict," *Annual Review of Fluid Mechanics* 30 (1998).

33  **embezzled funds and ran off to France:** E. Gold, "Lewis Fry Richardson, 1881–1953," *Obituary Notices of Fellows of the Royal Society* 9.1 (1954): 217–35.

34  **passage of Halley's Comet:** Peter Lynch, *The Emergence of Numerical Weather Prediction: Richardson's Dream* (Cambridge: Cambridge Univ. Press, 2006), 106.

34  **"heap of hay in a cold rest billet":** Richardson, *Weather Prediction by Numerical Process,* 219.

34  **"careful and conscientious driver":** Ashford, *Prophet—or Professor?,* 159.

37  **dispersion of chemical weapons:** Fleming, *Inventing Atmospheric Science,* 39.

38  **Norway had only nine weather observation stations:** Friedman, *Appropriating the Weather,* 121.

38  **photograph from the era:** See Friedman, *Appropriating the Weather,* 154; also https://www.uib.no/gfi/56744/bergensskolen-i-meteorologi.

39  **used in the forecast for the Allied invasion of Normandy:** Sverre Petterssen and James R. Fleming, *Weathering the Storm: Sverre Petterssen, the D-Day Forecast, and the Rise of Modern Meteorology* (Boston: American Meteorological Society, 2001), 209.

39  **"old and sadly":** Ibid., 29.

40  **"wrinkles in the face of Weather":** Quoted in Fleming, *Inventing Atmospheric Science,* 74.

40  **"It is ironic":** Frederik Nebeker, *Calculating the Weather: Meteorology in the 20th Century* (San Diego, CA: Academic Press, 1995), 57.

## CHAPTER 3: THE WEATHER ON EARTH

The World Meteorological Organization's online resources are remarkable; a look at the entire observing system is accessible via the OSCAR tool, with which I spent many hours for this chapter. I'm especially grateful for the guidance of John Zillman, of Australia's Bureau of Meteorology, and John Huntington in Brooklyn. Paul Sauer, a weather observer at LaGuardia Airport, let me tag along in his duties, while Jim Peters of the FAA happily arranged my visit. My journey to Utsira was met with an unusually gracious welcome: Atle Grimsby and Arnstein Eek allowed me Utsira's shortest ever "artist residency," and Anne Marthe Dyvi shared her vivid sense of the island. And of course Hans Van Kampen let me join him for an afternoon as a professional weather watcher on the edge of the world.

49  outline map of Norway: Friedman, *Appropriating the Weather*, 122.

54  sentimental notoriety: Charlie Connelly, *Attention All Shipping: A Journey Round the Shipping Forecast* (London: Abacus, 2005).

## CHAPTER 4: LOOKING DOWN

Harry Wexler is another fascinating meteorologist with an untold story, and my timing was fortunate in that I had the benefit once again of James Rodger Fleming's work, in this case his triple biography, *Inventing Atmospheric Science*. The themes of Paul Edwards's *A Vast Machine* pervade my own book, but my debt in this chapter is more direct. Edwards's argument for the globalness of weather infrastructure was a crucial insight that guided my own journey.

62  sailing observation ships: Shirlee Smith Matheson, *Amazing Flights and Flyers* (Calgary: Frontenac House, 2010), 65.

63  far enough north to be free: Alec Douglas, "The Nazi Weather Station in Labrador," *Canadian Geographic* (Dec. 1981/Jan. 1982).

66  then it began looking back: Clyde T. Holliday, "Seeing the Earth from 80 Miles Up," *National Geographic* (October 1950).

66  "so our forecasters can obtain a glimpse": Quoted in Angelina Long Callahan, "Satellite Meteorology in the Cold War Era: Scientific Coalitions and International Leadership 1946–1964" (PhD diss., Georgia Institute of Technology, 2013), 78.

67  "If guided missiles carrying cameras": Holliday, "Seeing the Earth from 80 Miles Up."

67  top-secret 1951 report: Stanley Greenfield and William Kellogg, "Inquiry into the Feasibility of Weather Reconnaissance from a Satellite Vehicle," RAND Report R-218 (April 1951).

67 "everlasting shortcoming": Jack Bjerknes, "Detailed Analysis of Synoptic Weather as Observed from Photographs Taken on Two Rocket Flights over White Sands, New Mexico, July 26, 1948," appendix to "Inquiry into the Feasibility of Weather Reconnaissance from a Satellite Vehicle," RAND Report R-218 (April 1951).

68 After joining the Weather Bureau: Fleming, *Inventing Atmospheric Science*, 136.

68 present at the Trinity atomic test: James R. Fleming, "Polar and Global Meteorology in the Career of Harry Wexler, 1933–62," in *Globalizing Polar Science: Reconsidering the International Polar and Geophysical Years*, edited by R. D. Launius, J. R. Fleming and D. H. DeVorkin (New York: Palgrave, 2010).

69 meteorology had been limited to two "eyepieces": Harry Wexler, "Structure of Hurricanes as Determined by Radar," *Annals of the New York Academy of Sciences* 48 (1947): 821–24, https://doi.org/10.1111/j.1749-6632.1947.tb38495.x, as quoted in Fleming, *Inventing Atmospheric Science*.

70 *Life* magazine treated the snapshot: "A 100 Mile High Portrait of Earth," *Life*, September 5, 1955.

70 "No one had suspected": Harry Wexler, "The Satellite and Meteorology," *Journal of Astronautics* 4 (Spring 1957).

70 painting he commissioned: James R. Fleming, "A 1954 Color Painting of Weather Systems as Viewed from a Future Satellite," *Bulletin of the American Meteorological Society* 88 (2007).

70 "Since the satellite will be": Wexler, "The Satellite and Meteorology."

71 encouraged Wexler to publish: Fleming, "A 1954 Color Painting."

71 "tiny unseen whirls and vortices": Harry Wexler, "Meteorology in the International Geophysical Year," *Scientific Monthly* 84 (1957).

71 "This global aspect of meteorology": Wexler, "The Satellite and Meteorology."

71 Wexler died of a heart attack: Fleming, *Inventing Atmospheric Science*, 189.

72 eighteen-sided drum the size of a breakfast table: Janice Hill, *Weather from Above: America's Meteorological Satellites* (Washington, D.C.: Smithsonian Institution Press, 1991), 11.

73 "earth doesn't look so big": Richard Witkin, "Vast Gains Seen for Forecasting," *New York Times*, April 1, 1960.

74 **"It comes down in rain":** Michael O'Brien, *John F. Kennedy: A Biography* (New York: Thomas Dunne, 2005), 894.

75 **sending into orbit cosmonaut Yuri Gagarin:** Edwards, *A Vast Machine*, 222.

75 **Wiesner commissioned the Norwegian meteorologist Sverre Petterssen:** Fleming, *Inventing Atmospheric Science*, 207.

78 **"a genuinely global infrastructure":** Edwards, *A Vast Machine*, 242.

78 **By 1975, one hundred weather bureaus:** Charles H. Vermillion and John C. Kamowski, "Weather Satellite Picture Receiving Stations, APT Digital Scan Converter," NASA Report TN D-7994, May 1975.

79 **overturning the historical understanding:** Callahan, "Satellite Meteorology in the Cold War Era," 3.

## CHAPTER 5: GOING AROUND

I was fortunate to attend the EUMETSAT Climate Symposium early on in my reporting, which gave a solid foundation for the broader state of meteorological satellites. At EUMETSAT itself, Kenneth Holmund, Yves Buhler, Nico Feldmann and Valerie Barthmann were remarkably transparent and forthcoming in helping me understand their admirable system. Tillmann Mohr's historical insight and fierce defense of the broad arc of weather satellites were crucial in shaping my thinking.

82 **constellation of overlapping eyes:** Tillmann Mohr, "The Global Satellite Observing System: A Success Story," WMO Bulletin vol. 59(1), January 2010.

## CHAPTER 6: BLASTING OFF

I am grateful to Dara Entekhabi of MIT, who answered my questions and encouraged my ongoing interest in each stage of SMAP's progress. At the Jet Propulsion Laboratory in Pasadena, Alan Buis, Sam Thurman and Simon Yueh walked me through the details of the spacecraft. And at Vandenberg Air Force Base, Tyrona Lawson and George Diller provided access to the pre-launch events.

93 **"tapping of the typewriter":** Jack Bjerknes, "Half a Century of Change in the 'Meteorological Scene,'" *Bulletin of the American Meteorological Society* 45 (1964).

98 developed a primitive guided missile: "Jet Propulsion Laboratory," NASA Facts, https://www.jpl.nasa.gov/news/fact_sheets/jpl.pdf.

101 "Copernican revolution": As quoted in Stephen Graham, *Vertical: The City from Satellites to Bunkers* (London: Verso, 2016), 29.

## CHAPTER 7: THE MOUNTAINTOP

By far, the greatest challenge of this book was learning about the weather models at the center of the weather machine, but thanks to the generosity and openness of the modeling community, that challenge was always conceptual and never logistical. My attendance at the annual WRF User's Workshop at the National Center for Atmospheric Research in Boulder was a crucial introduction to the topic, and I'm grateful in particular for conversations with Joe Klemp, George Bryan, Greg Thompson, Chris Davis, Rich Loft and Jeffrey Anderson. Hendrik Tolman at the National Centers for Environmental Prediction; Roland Potthast at the Deutscher Wetterdienst; and David Walters, Andrew Lorenc and Roger Saunders at the UK MET all furthered my understanding of models and how they fit into the global system of observation.

112 "sky is quite literally the limit": Stuart W. Leslie, " 'A Different Kind of Beauty': Scientific and Architectural Style in I. M. Pei's Mesa Laboratory and Louis Kahn's Salk Institute," *Hist. Stud. Nat. Sci.* 38.2 (2008): 173–221.

112 "monastic, ascetic, but hospitable": Lucy Warner, *The National Center for Atmospheric Research: An Architectural Masterpiece* (Boulder: University Corporation for Atmospheric Research, 1985), 13.

## CHAPTER 8: THE EURO

The European Centre for Medium-Range Weather Forecasts was the place that inspired this exploration, and my time there was undoubtedly its highlight. I am grateful to Dick Dee and Adrian Simmons for their considerable effort making my visit a success, and their warmth in welcoming me to Reading. The two of them, along with Peter Bauer, Tim Hewson, Florence Rabier, Alan Thorpe and Tim Palmer, are the world's finest, and I was grateful for the time they spent explaining their complex and crucial systems.

124 **emerged in the late 1960s:** Austin Woods, *Medium-Range Weather Prediction: The European Approach* (New York: Springer, 2006), 21.

125 **999-year ground lease:** Ibid., 13.

126 **ECMWF's scientists had squeezed out:** P. Bauer, A. Thorpe, and G. Brunet, "The Quiet Revolution of Numerical Weather Prediction," *Nature* 525.7567 (2015), https://doi.org/10.1038/nature14956.

## CHAPTER 9: THE APP

The reach of the Weather Company's system is remarkable, and I was grateful to Peter Neilley and his team for their willingness to walk me through it, especially Lea Armstrong and Jim Lidrbauch, and Campbell Watson for the helpful background. Joe Brown and Susan Murcko at *Popular Science* supported the reporting in this chapter, parts of which are adapted from my article, "This forecast brought to you by math" (*Popular Science*, July/August 2017). Jeff Masters is a legend, and it was wonderful to hear his story of Weather Underground's beginnings.

145 **that became everybody on the Internet:** Jeff Masters, "The Weather Underground Experience: 1991–2012," https://slideplayer.com/slide /219226/.

## CHAPTER 10: THE GOOD FORECAST

It is a contentious moment in weather forecasting, in the United States in particular, and I was grateful for thoughtful conversations with Kenny Blumenfeld, Bob Henson, and Eve Gruntfest. Early on in this project, Andrew Freedman, Eric Holthaus and Jason Samenow all offered their encouragement, and I benefited throughout from their incisive journalism. Louis Uccellini, Ryan Hanrahan and Dan Satterfield each contributed important insights into how the role of forecaster is changing.

160 **"Forecasts possess no":** Allan H. Murphy, "What Is a Good Forecast? An Essay on the Nature of Goodness in Weather Forecasting," *American Meteorological Society* 8 (June 1993).

## CHAPTER 11: THE WEATHER DIPLOMATS

In my effort to understand the workings of the WMO and the events of the congress, Bruce Angle of the Meteorological Service of Canada and Marjorie McGuirk were excellent early guides, pointing me in the right direction and offering introductions. I was grateful for the candid and precise comments of Courtney Draggon and Laura Furgione at the National Weather Service, Bruce Truscott and Andy Brown at the UK MET, and John Zillman of the Australian Bureau of Meteorology.

**165 bought it that week at Sparhawk's:** Susan Solomon, John S. Daniel and Daniel L. Druckenbrod, "Revolutionary Minds," *American Scientist* 95 (2007).

**167 "procure a tolerable house here":** Thomas Jefferson to Martha Jefferson Randolph, April 4, 1790, https://founders.archives.gov/documents /Jefferson/01-16-02-0172.

**167 "I should like to compare":** Thomas Jefferson to Thomas Mann Randolph, Jr., April 18, 1790, https://founders.archives.gov/docu ments/Jefferson/01-16-02-0202.

**167 Jefferson worked through the technical challenges:** Solomon, Daniel and Druckenbrod, "Revolutionary Minds."

**167 speculated about a broader observation system:** Edwin T. Martin, *Thomas Jefferson: Scientist* (New York: Collier Books, 1961), 124.

**175 "the most successful international system yet devised":** John W. Zillman, "Fifty Years of World Weather Watch: Origin, Implementation, Achievement, Challenge," *Bulletin of the Australian Meteographic and Oceanographic Society* 26 (2015).

# SELECTED BIBLIOGRAPHY

Ashford, Oliver M. *Prophet—or Professor?: The Life and Work of Lewis Fry Richardson*. Bristol, UK: Hilger, 1985.

Connelly, Charlie. *Attention All Shipping: A Journey Round the Shipping Forecast*. London: Abacus, 2005.

Cox, John D. *Storm Watchers: The Turbulent History of Weather Prediction from Franklin's Kite to El Niño*. New York: John Wiley, 2002.

Dyson, George. *Turing's Cathedral: The Origins of the Digital Universe*. New York: Vintage, 2012.

Edwards, Paul N. *A Vast Machine: Computer Models, Climate Data, and the Politics of Global Warming*. Cambridge, MA: MIT Press, 2010.

Fleming, James R. *Inventing Atmospheric Science: Bjerknes, Rossby, Wexler, and the Foundations of Modern Meteorology*. Cambridge, MA: MIT Press, 2016.

———. *Meteorology in America, 1800–1870*. Baltimore: Johns Hopkins University Press, 1990.

Friedman, Robert M. *Appropriating the Weather: Vilhelm Bjerknes and the Construction of Modern Meteorology*. Ithaca, NY: Cornell University Press, 1989.

Gleick, James. *The Information: A History, a Theory, a Flood.* New York: Vintage Books, 2012.

Graham, Stephen. *Vertical: The City from Satellites to Bunkers.* London: Verso, 2016.

Hill, Janice. *Weather from Above: America's Meteorological Satellites.* Washington, D.C.: Smithsonian Institution Press, 1991.

Lynch, Peter. *The Emergence of Numerical Weather Prediction: Richardson's Dream.* Cambridge: Cambridge Univ. Press, 2006.

Martin, Edwin T. *Thomas Jefferson: Scientist.* New York: Collier Books, 1961.

Matheson, Shirlee Smith. *Amazing Flights and Flyers.* Calgary: Frontenac House, 2010.

Monmonier, Mark. *Air Apparent: How Meteorologists Learned to Map, Predict, and Dramatize Weather.* Chicago: Univ. of Chicago Press, 1999.

Moore, Peter. *The Weather Experiment: The Pioneers Who Sought to See the Future.* New York: Farrar, Straus and Giroux, 2015.

Nebeker, Frederik. *Calculating the Weather: Meteorology in the 20th Century.* San Diego, CA: Academic Press, 1995.

Petterssen, Sverre, and James R. Fleming. *Weathering the Storm: Sverre Petterssen, the D-Day Forecast, and the Rise of Modern Meteorology.* Boston: American Meteorological Society, 2001.

Richardson, Lewis F. *Weather Prediction by Numerical Process.* Cambridge: Cambridge Univ. Press, 1922.

Sandlin, Lee. *Storm Kings: The Untold History of America's First Tornado Chasers.* New York: Pantheon Books, 2013.

Wilkinson, Alec. *The Ice Balloon: S. A. Andrée and the Heroic Age of Arctic Exploration.* New York: Vintage Books, 2013.

Woods, Austin. *Medium-Range Weather Prediction: The European Approach: The Story of the European Centre for Medium-Range Weather Forecasts.* New York: Springer, 2006.

# INDEX